Hanna M. Jacobson

July 1970

Fundamental
Research Statistics
for the
Behavioral Sciences

Fundamental
Research
Statistics

for the
Behavioral Sciences

John T. Roscoe
Kansas State University

HOLT, RINEHART AND WINSTON, INC.

New York · Chicago · San Francisco · Atlanta · Dallas
Montreal · Toronto · London · Sydney

TO THE STUDENTS AND FACULTY OF
COLORADO STATE COLLEGE

PREFACE

This text is intended to provide an introductory treatment of research statistics for the student in education and psychology. The content is sufficient to provide a rather substantial course of study for two terms; however, the method of organization permits the professor to edit the content liberally and should encourage its use in many courses scheduled for a single term. No background other than elementary high school algebra is assumed.

A major goal in writing the text was to achieve a level of readability appropriate for the beginning graduate student in education and psychology—something which the students tell us is not true of the typical textbook in this field. The approach to statistical theory emphasizes the conceptual rather than the mathematical. Students have been very vocal in their opposition to the traditional presentation of statistical theory, and it is hoped that some of this opposition may be overcome by the conceptual approach. A conscious effort has been extended to illustrate probability concepts with topics of interest to the student in education and psychology rather than the traditional coin flipping and drawing of marbles from urns.

The chapters have been kept very short, and they are more numerous than in the typical statistics textbook. Three benefits are expected from this type of organization: (1) Short learning units are notoriously more popular with students; (2) Professors may conveniently edit the text and control the content of their courses; and (3) The materials are conveniently classified for reference purposes.

One of the unique features of the text is the functional treatment of the various techniques of statistical inference. The emphasis is upon the selection of the most appropriate analysis for a given research situation. The most popular nonparametric procedures are presented in a fashion which encourages their consideration as alternatives to the parametric procedures. The statistical tables in the appendix, especially those for the nonparametric statistics, are superior for a text of this level.

Every statistical procedure is illustrated with a practical example of the type of thing encountered in elementary behavioral research. Many short, simple exercises are included, and an answer guide will be found in the appendix. The examples and the exercises utilize small numbers of small numbers—this is done because the author believes little insight into research analysis is gained from many hours of arithmetic labor. The student should note, however that most practical research does indeed involve large quantities of data and many extensive calculations. In our generation, these calculations will ordinarily be handled by an electronic computer.

The author is indebted to the Literary Executor of the late Sir Ronald A. Fisher, F.R.S., to Dr. Frank Yates, F.R.S., and to Oliver & Boyd Ltd., Edinburgh, for permission to abridge Tables 7 and 11 from their book *Statistical Tables for Biological, Agricultural and Medical Research*.

Finally, the author is much indebted to Drs. Jack N. Sparks and Beatrice Heimerl who reviewed the manuscript and offered many helpful suggestions. They should not be held responsible for any inadequacies in the text. A note of thanks is also extended to Marvin H. Halldorson, Timothy H. Hoyt, Gary C. Stock, Eric R. Strohmeyer, and David E. Suddick, who were graduate students at the time of writing and who prepared the answer guide found in the Appendix.

Manhattan, Kansas   J. T. R.
March 1969

CONTENTS

1

INTRODUCTION TO STATISTICS

1.1 STATISTICS

Statistics, as a modern academic discipline, provides scientific procedures for collecting, organizing, summarizing, and analyzing quantitative information of the type commonly encountered in the behavioral sciences. First, statistics permits the summarization and presentation of large quantities of this information in such a fashion as to facilitate its communication and interpretation. Second, statistics enables the scientist to extend his research far beyond the restricted setting in which most research is actually conducted. It is this latter application, the use of statistics as a tool for analysis and extension of research findings, with which we are primarily concerned.

Some students and practitioners in the behavioral sciences shy away from quantitative information in general and statistics in particular. Perhaps through ignorance of the discipline, fear of its rigors, or philosophic bias, they have chosen instead to draw their conclusions from tradition, authority, or intuition. Such persons, of course, should not be classed as behavioral scientists. Reliable evidence from empirical research invariably lends itself to quantification and subsequent statistical analysis.

The use of statistical procedures in empirical research is not limited to the behavioral sciences (education, psychology and sociology, for instance). The use of these procedures in research is now universal in such diverse fields as agriculture, biology, chemistry, economics, medicine, and physics. For example, statistical procedures enable the educator to draw conclusions with respect to the efficacy of various instructional methods and materials, the psychologist to determine the precision with which he measures certain human traits, the sociologist to speak with confidence about the incidence of antisocial behavior, the physicist to interpret the activities of subatomic particles, the medical scientist to choose the most effective medicine, and the agricultural engineer the most productive fertilizer.

1.2 DESCRIPTIVE STATISTICS

The techniques of *descriptive statistics* enable us to describe with precision a collection of quantitative information, to do this in a form that is more concise and convenient than the original collection, and to do it in a fashion that makes for ease of interpretation. For instance, an educational researcher might choose to summarize a large number of test scores from a specified population of students by calculating the average score. His description might be further enriched by calculating a measure of dispersion—a single numerical value that reveals valuable information about the manner in which the scores are distributed. Or, he might choose to summarize the collection of scores with a graphical representation. Each of these techniques of descriptive statistics is intended to facilitate the orderly communication and interpretation of an otherwise disorganized mass of raw data.

1.3 STATISTICAL INFERENCE

The techniques of *statistical inference* enable the researcher to draw inferences and generalizations from small groups, called *samples*, to larger groups, called *populations*—from individuals who are direct participants in experimental research to individuals who are not participants—and to do so with a well-defined degree of confidence. A sample is a relatively small group of individuals chosen in a scientific fashion to represent a relatively large group of individuals (the population) which the researcher is interested in studying. The student should take particular note of the fact that the researcher is primarily interested in the population although his research activities are carried on with the sample. This is an essential economy in research—that the scientist be able to extend with confidence his research findings beyond the narrow confines in which the research is conducted. The techniques of statistical inference provide the vehicle for scientific generalizations of this sort.

Statistical inference has been described as a collection of tools for making the best possible decisions in the face of uncertainty—that is, the situation where some, but not all, of the facts are in. Behavioral scientists have long been reconciled to making decisions of this type, and there is little prospect of their being delivered from this uncertainty.

1.4 PREDICTION

The terms *correlation, regression,* and *prediction* are so closely related in statistics that they are often interchangeable. These terms are used only for situations in which the researcher has at least two measures, one related to the other, on each individual. Related measures are those that tend to vary together—when one is larger, the other tends to be systematically larger or smaller. For example, height and weight of human beings tend to be positively correlated; as a general rule, taller persons are heavier persons. Thus, one might predict with a reasonable amount of confidence that a youngster who grows to be six feet tall will be heavier than one who grows to be five feet tall. When two measures are correlated, it is possible, though often with a limited degree of success, to predict one measure from the other. Thus, because IQ and academic success are usually correlated, an educator can predict the possible success of a student whose IQ is known with greater precision than he can predict the success of a student whose IQ is not known. Regression analysis permits prediction with a specified degree of precision.

In recent years, some behavioral scientists have been using regression analysis as a general procedure for experimental analysis. When an electronic computer is available, regression analysis provides an extremely flexible approach to very complex research problems. This text presents an introduction to these techniques along with the more traditional techniques of descriptive statistics and statistical inference.

INTRODUCTION TO MEASUREMENT

2.1 VARIABLES AND CONSTANTS

The scientific study of human behavior is based upon observations of that behavior. In order for the observations made by behavioral scientists to be meaningful, these observations must indicate variation from time to time or person to person. The term *sex*, for instance, would be meaningless if all persons were of the same sex. Similarly, the expression *intelligence* would also be meaningless if all persons had the same amount of this trait. Traits such as these, which are capable of variation from person to person, are called *variables*. The systematic study of the relationships between these variables is the subject matter of the behavioral sciences.

There are two broad classes of variables: those that vary in *quality* and those that vary in *quantity*. Of the examples given, sex is a qualitative variable, and intelligence is a quantitative variable.

Whenever a behavioral scientist undertakes the study of a group of individuals (a population), it will be necessary to describe that population in terms of characteristics that they have in common, that is, characteristics that do not vary from person to person in this particular group. For example, the population under study might be all female, all third graders, or all subject to the same experimental treatment. Any property that is the same for all members of the group is called a *constant*.

2.2 DISCRETE AND CONTINUOUS VARIABLES

If a variable can take on only a finite set of values, it is said to be *discrete*. When we say that a variable can take on only a finite set of values, this usually means that fractional values are not allowed. Such variables as sex, political affiliation, and number of children in a classroom are discrete variables.

If a variable can take on *any* value, including fractional values, over a range of values, it can take on an infinite set of values. Such a variable is called a *continuous* variable. Variables like height, weight, spelling ability, and IQ are continuous variables. Often, measures of this sort are reported as whole numbers, but this is due to imprecision of measurement rather than any restriction on the value of the trait. For instance, IQ scores are customarily reported as whole numbers, but this does not mean that a person could not have an IQ of $115\frac{1}{4}$. It simply means that an IQ of 115 is an approximation of some score between $114\frac{1}{2}$ and $115\frac{1}{2}$.

A continuous variable will always be a quantitative one, but a discrete variable may be either qualitative or quantitative. Family size is an example of a quantitative discrete variable, and sex is an example of a qualitative discrete variable.

Whether a variable is continuous or discrete will often determine the method of statistical analysis, since some statistical procedures are appropriate for use with continuous variables and others for discrete variables. Occasionally, it will be convenient to treat a continuous variable as though it were discrete. For example, IQ scores might be used to classify students as gifted, normal, or retarded. This reduction of a continuous scale of values to three categories might be useful in a given situation, but the researcher is more likely to regard it as wasteful of valuable information.

2.3 MEASUREMENT

We usually think of *measurement* as the assignment of numbers indicative of quantity to some object or trait that we seek to assess. Generally, it will be desirable (and often taken for granted) that the numbers so assigned will lend themselves to the arithmetic operations of addition, subtraction, multiplication, and division. This is not always the case, however, and it will be convenient for the behavioral scientist to think in terms of several different levels of measurement. The various levels of measurement are often called *measurement scales*.

2.4 NOMINAL MEASUREMENT

The lowest level of measurement is *nominal* or classificatory measurement, which consists simply of classifying observations into categories

These categories must be mutually exclusive and collectively exhaustive. Each observation must be capable of classification into one and only one category (there must not be any observations that cannot be assigned to one of the categories, and there must not be any observations that can be assigned to more than one category). For example, sex is a nominal variable, and we usually assume that all persons can be accurately classified as either male or female, and that no person can be classified as both. Similarly, all persons might be classified as college graduates or noncollege graduates, married or unmarried, etc. If the population were limited to secondary school teachers, these might be classified as teachers of English, mathematics, social studies, science, etc. Similarly, students might be classified on a nominal scale by state of birth or school attended.

Nominal scales are both discrete and qualitative. Occasionally, numbers will be assigned to nominal variables (students might be assigned to Group I, Group II, or Group III, for instance), but arithematic operations performed with these numbers are meaningless.

Any nominal variable may be *dichotomized*. For example, secondary school teachers can all be classified as either English teachers or teachers of other subjects. Students in college may be classified as residents of the state in which the college is located or as nonresidents of that state. The investigator may then treat these as *binary variables*, assigning a score of one to the members of one group and a score of zero to the members of the other group. Mathematical operations with these binary variables have proved to be extremely useful in the behavioral sciences, especially in regression analysis.

As a general rule, statistical procedures designed especially for use with nominal variables may also be used with higher order variables (the higher order measures may be treated as nominal measures), but this is not usually the way to get the most mileage out of the available data.

2.5 ORDINAL MEASUREMENT

An *ordinal* scale is distinguished from a nominal scale by the additional property of order among the categories: A category will be thought of as higher than or lower than the adjacent category. While the idea of relative quantity of a given trait is characteristic of an ordinal scale, nothing is specified with respect to the magnitude of the interval between two measures. For example, a professional musician can *rank* three pianists in order of their ability to perform a particular composition, but he cannot speak precisely of how much more talent one has over another. Similarly, a clinical psychologist might speak with confidence regarding a client's improvement, but it is unlikely that he could quantify the amount of improvement. Ordinal measurement may be likened to a foot race without a stop watch; the order in which the runners finish is determined, but the magnitude of the intervals between them is indeterminate. Ordinal measurement is quite common in the behavioral sciences.

Ordinal scales may be either discrete or continuous. For instance, teachers might be classified on a discrete ordinal scale as elementary, secondary, or college teachers. However, if three teachers were ranked from one to three on the basis of their contribution to the profession, a continuous ordinal scale is implied.

As a general rule, statistical procedures developed especially for use with ordinal measures require that the measures be continuously distributed. Discrete ordinal measures will usually be treated with statistical procedures designed for use with nominal variables. Statistical techniques which have been developed especially for use with nominal and ordinal measures are often called *nonparametric* statistics.

There is some disagreement as to whether ordinal variables should be classified as qualitative or quantitative. The use of arithmetic operations (addition, subtraction, multiplication, division) with the measures yielded by ordinal scales is questionable from a mathematical point of view but often useful from a practical point of view. Ordinal scales are sometimes called *rank* order scales.

2.6 INTERVAL MEASUREMENT

An *interval* type of scale is distinguished from an ordinal scale by having equal intervals between the units of measure. This means that a score of 50 points is halfway between scores of 40 and 60 points. The distinct advantage of such a scale is that the measures yielded may be treated with the common arithmetic operations of addition, subtraction, multiplication, and division. The interval scale is a truly quantitative scale.

One notable shortcoming of the interval scale is that it lacks a true zero. This means that one cannot interpret a score of 50 as indicating twice as much of a given trait as a score of 25. The problem arises from the fact that one cannot, as a general rule, meaningfully establish that a person has zero quantity of a particular psychological trait. For example, intelligence tests have been carefully scaled to yield interval measures, but there is no provision in the available tests to identify a person of zero intelligence. Similarly, a student may score zero points on an achievement test, but it has not been demonstrated that he has zero quantity of whatever trait the test is supposed to measure.

Most of the well-standardized psychological tests are intended to yield interval measures. The tests constructed by classroom teachers should probably be regarded as yielding ordinal measures, but no great harm seems to come from treating them as interval measures. While the data from interval scales may be analyzed using procedures designed for nominal and ordinal data, it is usually more profitable to take advantage of the higher level of measurement when it is available.

2.7 RATIO SCALES

Ratio scales have all the properties of interval scales, with the additional property of a true zero. While interval and ordinal scales are more often encountered by the behavioral scientist, he will also make use of such measures as age, height, and weight, which are measured on ratio scales.

With a ratio scale, one can intelligently speak of an individual having zero quantity of a particular trait, or of one individual having three times as much of this trait as another individual. While we may say that a person who weighs 150 pounds has twice the weight of a person who weighs 75 pounds, we cannot say that a person who has an IQ of 150 has twice the intelligence of a person with an IQ of 75.

The data yielded by ratio scales will usually be analyzed in the same fashion as data yielded by interval scales, and the statistical procedures which are specified for use only with interval and ratio data are customarily referred to as *parametric* statistics.

FREQUENCY DISTRIBUTIONS

3.1 INTRODUCTION

The behavioral scientist frequently encounters large masses of numbers, usually test scores, which communicate little meaning until these are organized or summarized with one of the techniques of descriptive statistics. Table 3.1, for instance, contains scores yielded by an English usage test administered to 120 college students. The scores were recorded in the order which they were received from the students.

Table 3.1 Random Arrangement of 120 Scores

13	7	6	16	9	3	12	13	11	14	5	10	4	9	7
9	17	14	13	11	12	19	4	9	7	12	16	7	12	11
4	5	11	17	10	9	10	17	13	16	12	13	18	9	15
16	12	9	7	14	3	16	9	5	10	17	15	11	10	10
6	18	10	15	8	13	14	18	12	8	9	8	12	8	12
11	8	11	12	16	11	7	15	11	14	20	7	6	14	16
5	15	19	14	13	15	8	8	13	11	7	11	12	17	13
14	10	13	8	5	17	11	10	6	12	5	1	11	8	10

It is extremely difficult to make any generalizations about the characteristics of this group of students from the collection of unordered scores. A little more meaning is communicated when the scores are arranged in numerical order, as in Table 3.2.

As indicated earlier, the techniques of descriptive statistics enable the researcher to describe with precision a collection of quantitative informa-

Table 3.2   Ordered Arrangement of 120 Scores

20	17	16	15	14	13	12	11	11	10	9	8	7	6	5
19	17	16	14	13	13	12	11	11	10	9	8	7	6	5
19	17	16	14	13	13	12	11	11	10	9	8	7	6	4
18	17	15	14	13	12	12	11	11	10	9	8	7	6	4
18	16	15	14	13	12	12	11	10	10	9	8	7	5	4
18	16	15	14	13	12	12	11	10	10	9	8	7	5	3
17	16	15	14	13	12	12	11	10	9	9	8	7	5	3
17	16	15	14	13	12	11	11	10	9	8	8	7	5	1

tion, to do this in a form that is more concise and convenient than the original collection, and to do it in a fashion that makes for ease of interpretation and communication. All of this can be accomplished without loss of information by arranging the data at hand in a *frequency distribution*.

3.2   FREQUENCY DISTRIBUTIONS WITH UNGROUPED DATA

A frequency distribution is a technique for systematically arranging a collection of measures on a given variable to indicate the frequency of occurrence of the different values of the variables. Consider Table 3.3, which contains all of the information of Tables 3.1 and 3.2, but in the form of a frequency distribution. The column marked X contains all possible score values, and the column marked f contains the score frequencies. The *frequency* of any score is simply the number of persons who received that score.

It is immediately apparent that the frequency distribution markedly facilitates interpretation of the collection of scores. Notice that the scores of highest frequency stand out clearly. The investigator is able to discern at a glance the general distribution of the scores along the score scale.

The usual procedure for making up a frequency distribution is first to list in order of size all possible score values from the highest to the lowest, and then to read off the scores in the collection in the order in which they were received, making a tally mark for each of these opposite the corresponding score value in the list. Table 3.4 illustrates this procedure for a portion of the data in the previous tables.

3.3   FREQUENCY DISTRIBUTIONS WITH GROUPED DATA

Sometimes the information contained in a frequency distribution will be appreciably less cumbersome and more redaily comprehended if the scores

Table 3.3 Frequency Distribution of 120 Ungrouped Scores	
X	f
20	1
19	2
18	3
17	6
16	7
15	6
14	8
13	10
12	12
11	13
10	10
9	9
8	9
7	8
6	4
5	6
4	3
3	2
2	0
1	1
0	0

Table 3.4 Illustration of the Procedure for Making a Frequency Distribution	
X	Tally marks
20	/
19	//
18	///
17	ЖИ /
16	ЖИ //

are grouped into intervals, each of which contains several score levels. Consider Table 3.5, in which each interval contains two score levels, and Table 3.6, in which each interval contains four score levels. The data are the same as contained in the previous tables.

Notice that Table 3.5 is more compact than any of the previous tables; however, the identity of the individual score is lost, and this loss of information may be of greater importance than the gain in compactness. Table 3.6 is even more compact, but at the expense of a loss of even more information. While Table 3.5 expresses the general form of the distribution quite well, Table 3.6 has lost the form in the grouping of the data.

Some older statistics textbooks suggest the grouping of interval or ratio scale measures to simplify calculations with that data. Modern behavioral scientists, however, tend to regard this as a great waste of research information. Generally, the value of the original data is much greater than any possible saving in calculation effort. Normally, the cal-

Table 3.5 Frequency
Distribution of 120
Grouped Scores

X	f
19–20	3
17–18	9
15–16	13
13–14	18
11–12	25
9–10	19
7–8	17
5–6	10
3–4	5
1–2	1

Table 3.6 Frequency
Distribution of 120
Grouped Scores

X	f
17–20	12
13–16	31
9–12	44
5–8	27
1–4	6

culations for any substantial research effort will be performed with an electronic computer, and grouping does nothing to facilitate the process.

However, the grouping of data in the fashion described may be a useful technique in descriptive statistics, where the frequency distribution is an end product rather than an intermediate step in a series of calculations. If the grouping of data is undertaken to facilitate interpretation, it is recommended that the number of intervals be between 10 and 25, with 15 to 20 being optimum. Of course, all intervals will normally contain the same number of score levels, with the possible exception of the top and bottom intervals, which the investigator may choose to leave open-ended to accommodate a few extreme scores. Some authors recommend that the interval contain an odd number of score levels so that the middle score is an integer and may be conveniently used as an index value representative of all measures in the interval.

3.4 CLASS LIMITS

Whenever measurements on continuously distributed traits are recorded as integers (whole numbers only, no fractions), a given score must be regarded as being somewhere between half a unit above or below the integer. For example, a youngster getting twelve words right on a spelling test must be regarded as having a score somewhere between $11\frac{1}{2}$ and $12\frac{1}{2}$, since spelling ability is believed to be a continuous rather than a discrete variable. The fact that the score is recorded in whole numbers only is a function of the precision of measurement rather than the discreteness of the variable. For this reason, the *real limits* of continuous data recorded with unit intervals will be regarded as extending one-half unit on either side of the recorded score.

The same principle applies to grouped data. If an interval contains scores ranging from 16 to 20, for instance, the real limits of the interval are $15\frac{1}{2}$ to $20\frac{1}{2}$. Occasionally, the term *integer limits* will be used; in the example quoted, the integer limits are 16 and 20.

The student will want to take note of the idea of real limits, since this is a useful concept which will appear repeatedly in the study of statistics.

3.5 CUMULATIVE AND RELATIVE FREQUENCIES

It is not unusual to report interval frequencies as decimal fractions, or percentages, of the total number of scores in the collection. These are known as *relative frequencies* (*rf*), and they may be simply calculated by the formula

$$rf = \frac{f}{N}$$

where *rf* is the relative frequency, f is the frequency of the interval, and N is the total number of scores in the collection. The sum of all the relative frequencies in the distribution must equal 1.00 or 100 percent. Relative frequencies are especially important in statistical inference, as they are intimately related to the mathematical idea of probability.

The *cumulative frequency* (*cf*) of a given interval is the frequency of the interval plus the total of the frequencies of all intervals below the given interval. Another way of stating this is to say that the cumulative frequency of any interval is the number of scores in the distribution which fall below the upper real limit of that interval. This is another important idea which will come up again in the study of statistical inference.

The *relative cumulative frequency* (*rcf*) of a given interval is the cumulative frequency divided by the number of measures in the collection:

$$rcf = \frac{cf}{N}$$

Table 3.7 contains the relative frequencies, cumulative frequencies, and relative cumulative frequencies for the ungrouped frequency distribution considered earlier. It also contains a column for fX, the product of the

Table 3.7   Frequency Distribution of 120 Scores

X	f	fX	cf	rf	rcf
20	1	20	120	.008	1.000
19	2	38	119	.017	.992
18	3	54	117	.025	.975
17	6	102	114	.050	.950
16	7	112	108	.058	.900
15	6	90	101	.050	.842
14	8	112	95	.067	.792
13	10	130	87	.083	.725
12	12	144	77	.100	.641
11	13	143	65	.108	.541
10	10	100	52	.083	.433
9	9	81	42	.075	.350
8	9	72	33	.075	.275
7	8	56	24	.067	.200
6	4	24	16	.033	.133
5	6	30	12	.050	.100
4	3	12	6	.025	.050
3	2	6	3	.017	.025
2	0	0	1	.000	.008
1	1	1	1	.008	.008
0	0	0	0	.000	.000

$$N = 120 \qquad T = 1327$$

interval score and frequency, which facilitates the calculation of the sum of the scores, which we will call T.

3.6   BIVARIATE FREQUENCY DISTRIBUTIONS

The concept of frequency distribution may be readily extended to the situation in which two measures are available on each subject, and all of this information may be reported in a single table. Such a table is called a *bivariate frequency table* or *contingency table*.

Suppose that the 120 scores reported in the previous univariate tables were accompanied by information which also identified the student as a freshman, sophomore, junior, or senior. It would be a simple matter to construct four different frequency distributions, one for each class, and then to place them side by side for comparison. All of this can be accomplished, however, with a single bivariate frequency distribution. Table 3.8

Table 3.8   Bivariate Frequency Distribution

Scores	1–4	5–8	9–12	13–16	17–20	Totals
Seniors	0	1	6	15	8	30
Juniors	0	3	9	14	4	30
Sophomores	3	7	18	2	0	30
Freshmen	3	16	11	0	0	30
Totals	6	27	44	31	12	120

makes use of the grouped data of Table 3.6, as well as knowledge of the class of each respondent, in a bivariate frequency table. Notice that the column sums are the same as the frequencies reported in Table 3.6, and that the row sums indicate that the sample was evenly divided among the four classes. These row and column sums are called *marginal frequencies.*

EXERCISES

3.1   Given the following collection of 100 IQ scores:

100	107	85	119	93	111	89	90	113	99
92	108	103	98	86	73	102	100	74	103 ·
112	97	115	72	104	117	98	127	89	107
126	99	71	99	85	95	121	84	96	83
101	66	123	107	97	101	86	106	114	100
85	110	71	102	118	111	103	83	99	126
102	98	96	130	88	114	100	100	92	100
96	81	122	92	80	91	119	81	102	93
118	92	98	88	112	95	110	136	97	98
106	112	76	113	77	100	97	105	101	89

   (a)   Construct a frequency distribution using unit intervals.
   (b)   Construct a frequency distribution using intervals of five points. Let 65–69 be the lowest interval.
   (c)   Construct a frequency distribution using intervals of ten points. Let 60–69 be the lowest interval.
   (d)   Which of the three frequency distributions communicates the general distribution of the IQ scores most effectively? Which of the three distributions contains the most information?

3.2   Given the following collection of scores on a spelling test: 5, 6, 10, 8, 8, 4, 5, 5, 9, 8, 6, 6, 7, 4, 5, 5, 5, 3, 6, 9
   (a)   Construct a frequency distribution using unit intervals.
   (b)   Determine the values of N and T.

(c) Determine the cumulative frequency for each interval.

(d) Determine the relative frequency for each interval.

(e) Determine the relative cumulative frequency for each interval.

3.3 In a recent study of high school student attitudes toward a certain phase of American foreign policy, 22 boys indicated agreement with the policy, 44 indicated disagreement, and 34 had no opinion, while 62 girls indicated agreement, 27 indicated disagreement, and 12 had no opinion. Construct a bivariate frequency distribution with this data.

PERCENTILE RANKS

4.1 SCORE TRANSFORMATIONS

The numerical value of a student's score on a test is a highly arbitrary thing. It is a function of the number of items in the test and of the difficulty of the items. The interpretation of the score may be greatly enriched by expressing it in units which are a function of its relationship to other scores in the distribution. A score has meaning only as it is related to other scores, which is another way of saying that the principal interest of the behavioral sciences is the way in which individuals differ.

When a score in a collection of such scores is expressed as the percentage of scores in the collection that are below this score, the percentage is called a *percentile rank*. A percentile rank is a relative rank, a rank order score based on a scale of 100.

Translating a score into a percentile rank in the manner described is one of several possible *score transformations* that are intended to convey the relationship of a given score to the group. Score transformations may also be used to change the shape of a frequency distribution or to facilitate the comparison of scores from different instruments. When a score is transformed to a new score scale, the original score is referred to as the *raw* score and the new score as the *transformed* score.

To report that a student scored 40 points on an English test and 35 points on a mathematics test has little meaning of itself, but to report that he scored at the 40th percentile in English and at the 60th percentile in mathematics immediately conveys the idea that the student scored slightly below the group average in English and slightly above in mathematics. Such a comparison implies that the two percentile ranks were calculated with reference to a single group or two equivalent groups. Of course, it is important that the group to which the individual is compared be well defined.

4.2 DEFINITIONS

The *percentile rank* of a given raw score is the percentage of scores which is below this particular score. For example, if 50 percent of the students in a distribution are below a given student, his percentile rank is 50. Notice that the percentile rank is a point on the *tranformed* score scale. The abbreviation PR will be used to represent percentile rank.

A *centile* (or percentile as it is called by some authors) is a point on the *raw* score scale which corresponds to a given percentile rank. For example, an IQ score of 100 ordinarily corresponds to the 50th centile. The percentile rank is 50, the 50th centile is 100. There are 99 centiles, and these divide a frequency distribution into 100 equal parts. The letter C with a subscript will be used to represent centiles, the subscript indicating which centile. The symbol C_{50}, for instance, represents the 50th centile.

There are nine *deciles*, and these divide a frequency distribution into ten equal parts. Thus, the 1st decile (D_1) is the 10th centile (C_{10}), and the 2d decile (D_2) is the 20th centile $C_{(20)}$, and so forth.

There are three *quartiles*, and these divide a frequency distribution into four equal parts. Thus, the 1st quartile (Q_1) corresponds to the 25th centile (C_{25}), the 2d quartile (Q_2) to the 50th centile (C_{50}), and the 3d quartile (Q_3) to the 75th centile (C_{75}). The 2d quartile is more often referred to as the *median*. The student is reminded that a quartile is a *point* on a score scale rather than a portion of the distribution.

4.3 CALCULATION OF PERCENTILE RANKS

The percentile rank of any score in a frequency distribution with unit intervals may be calculated by the formula:

$$PR = \frac{100}{N}\left(cf - \frac{f}{2}\right)$$

where f is the frequency of the score, cf is the cumulative frequency, and N is the number of scores in the entire distribution. The expression inside the parentheses is the cumulative frequency at the midpoint of the interval (cfm) in which the score appears. Dividing the expression by N produces the relative cumulative frequency at the midpoint $(rcfm)$, and multiplying by 100 moves the decimal to the right two places to produce a percentage from a proportion. It is conventional to report percentile ranks rounded to the nearest whole number.

Table 4.1 illustrates the calculation of percentile ranks for the frequency distribution that was used for illustrative purposes in Chap. 3. The actual

Table 4.1   Percentile Ranks of 120 Scores

X	f	cf	cfm	rcfm	PR
20	1	120	119.5	.996	99.6
19	2	119	118	.983	98.3
18	3	117	115.5	.962	96.2
17	6	114	111	.925	92.5
16	7	108	104.5	.871	87.1
15	6	101	98	.817	81.7
14	8	95	91	.758	75.8
13	10	87	82	.683	68.3
12	12	77	71	.592	59.2
11	13	65	58.5	.487	48.7
10	10	52	47	.392	39.2
9	9	42	37.5	.312	31.2
8	9	33	27.5	.229	22.9
7	8	24	20	.167	16.7
6	4	16	14	.117	11.7
5	6	12	9	.750	7.5
4	3	6	4.5	.037	3.7
3	2	3	2	.017	1.7
2	0	1	1	.008	0.8
1	1	1	0.5	.004	0.4

calculations used in determining the percentile rank corresponding to a raw score of 17 are as follows:

$$PR = \frac{100}{120}\left(114\tfrac{6}{2}\right) = \frac{100}{120}(111) = 92.5$$

The student should take note of the fact that the raw scores from which percentile ranks are calculated may be ordinal, interval, or ratio data; however, the transformed scores are ordinally scaled. This suggests that adding two percentile ranks or performing similar arithmetic operations with them is mathematically questionable. However, in actual practice, the use of arithmetic operations with percentile ranks is not at all uncommon and appears to work out quite satisfactorily in many circumstances.

Some authors give formulas for calculating percentile ranks from grouped data, but this is likely to yield a transformed score distribution with many gaps in it, and the practice is not recommended. Percentile ranks are most appropriate when calculated from ungrouped raw scores, from a distribution containing at least 200 scores.

4.4 THE DISTRIBUTION OF PERCENTILE RANKS

The distribution of percentile ranks is uniform throughout the transformed score scale. Theoretically, there will be an equal number of scores (one-fifth of the total) below the 20th centile, between the 20th and 40th centiles, between the 40th and 60th, between the 60th and 80th, and above the 80th. For a given collection of scores, this may not be exactly the case, especially if the number of scores is small; however, if the number of scores is several hundred or more, and the raw score scale is relatively continuous (having many levels of performance), the distribution will be almost perfectly uniform. When a graph is drawn of such a distribution using the techniques discussed in the following chapter, the graph will be rectangular in shape. The distribution of percentile ranks is said to be rectangular or uniform.

4.5 THE CALCULATION OF CENTILES, DECILES, AND QUARTILES

If the student will refer to the PR column in Table 4.1, he will find that none of the raw scores in this distribution corresponds precisely to any centile. For instance, the median (which is also C_{50}, D_5, and Q_2) lies somewhere between a score of 11 and a score of 12. When a centile does not correspond directly with any recorded score, the following procedure is recommended for determining its value:

(1) Let $p = $ PR/100 where PR is the percentile rank corresponding to the desired centile. For example, $p = .25$ for Q_1 and $.75$ for Q_3. Multiply p by N to get Np.

(2) If the cumulative frequency of any interval is equal to Np, the desired centile is equal to the upper real limit of the interval. Referring to Table 4.1, one may use this procedure to determine that C_5 is 4.5, C_{10} is 5.5, and C_{20} is 7.5. However, the bulk of the centiles are still indeterminate, and we must proceed to the next step for these.

(3)   Determine the interval which has the smallest cumulative frequency larger than Np. Then calculate the desired centile by the following formula:

$$C = X_u - \frac{cf_i - Np}{f_i}$$

where X_u is the upper real limit, f_i is the frequency, and cf_i is the cumulative frequency of the interval.

Now, let us illustrate the use of this procedure for determining the values of Q_1, Q_2, and Q_3.

$$Q_1 = 8.5 - \frac{33 - 30}{9} = 8.17$$

$$Q_2 = 11.5 - \frac{65 - 60}{13} = 11.12$$

$$Q_3 = 14.5 - \frac{95 - 90}{8} = 13.88$$

A graphic method for determining the values of the various centiles will be considered in Chap. 5.

EXERCISES

4.1   Refer to the data for Exercise 3.1. Calculate the percentile rank for each score in the distribution. Determine the values of Q_1, Q_2, and Q_3.

4.2   Refer to the data for Exercise 3.2. Calculate the percentile rank for each score in the distribution. Determine the values of D_2, D_4, D_6, and D_8.

5

GRAPHS OF FREQUENCY DISTRIBUTIONS

5.1 INTRODUCTION

One of the techniques which may be employed to organize and summarize a collection of test scores is to represent the data graphically. The graphs which are described in this chapter correspond directly to the frequency distributions of Chap. 3 and often perform the same functions, communicating the same information pictorially. As the student progresses into the study of statistical inference, he will find that graphs not only portray the form of a frequency distribution but also facilitate the solution of many problems in research.

All of the graphs which will be considered here will be constructed with reference to two perpendicular lines, one horizontal and one vertical, which are called the *coordinate axes*. The horizontal axis (or *abcissa*) will represent the score scale, much in the fashion that a ruler is used. If the data are grouped, the investigator will probably choose to plot the mid-points of the intervals along this axis. Of course, this will be much simpler if the intervals are chosen so that the midpoint is an integer. Frequencies or relative frequencies will be plotted along the vertical axis (or *ordinate*). It is important in the construction of any graph that both axes be clearly labeled so that the reader knows what they represent. The graph, itself, should also be named or labeled in such a fashion that the reader will know what it is all about.

One of the important ideas associated with graphs of frequency distributions is that the *area* under a graph (or portion of a graph) is representative of the frequency of the corresponding scores. Often, the total area under such a graph will be set to unity, in which case the area above a certain portion of the score scale will be equal to the relative frequency of those scores. This relationship between relative frequency and area is basic to the use of statistics in research.

5.2 THE HISTOGRAM

A *histogram* is essentially a bar graph of a frequency distribution in which the frequencies are represented by areas in the form of vertical bars. Its purpose is to show the frequencies of the various scores graphically. It carries precisely the same information as the original frequency distribution from which it was made, although the graph becomes only an approximation of the frequency distribution if the number of scores is large.

Figures 5.1, 5.2, and 5.3 are histograms of the frequency distributions of Tables 3.3, 3.5, and 3.6 respectively. Notice that the width of each bar stretches from the lower real limit of the score interval to the upper real limit.

In the construction of a histogram, the score scale may be laid out by labeling either the midpoint or the limits of the intervals. The use of the midpoint is the more common practice. If the scores do not range down to

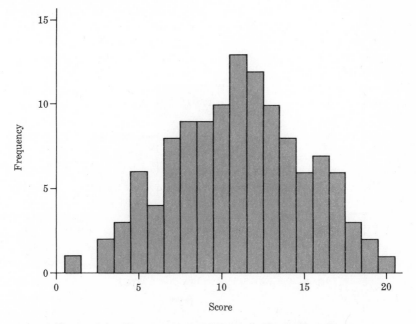

Figure 5.1 Histogram of 120 English Usage Test Scores.

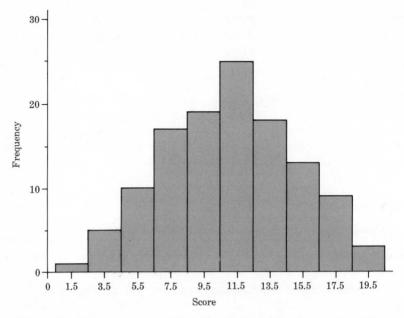

Figure 5.2   Histogram of 120 Scores Grouped in Intervals of Two Points.

Figure 5.3   Histogram of 120 Scores Grouped in Intervals of Four Points.

zero, it is permissible to start the score scale at some convenient point above zero which makes provision for the lowest score. Empty score intervals between the highest and lowest score values should be shown on the graph (as in Fig. 5.1), and it is a good practice to include one empty interval at each end of the histogram.

The frequency scale must be laid out to provide for all frequencies from zero to the largest frequency if the relationship between frequency and area is to be maintained. The form of the histogram will be the same whether the vertical axis is laid out with frequencies or relative frequencies.

5.3 THE FREQUENCY POLYGON

The frequency distribution may also be portrayed by means of a frequency *polygon*, a graph consisting of straight lines connecting points located above the midpoints of the intervals at heights corresponding to the frequencies. An empty interval is included at each end of the distribution, and the graph is brought down to the horizontal axis (which corresponds to zero frequency) at the midpoints of these two intervals. When the graph is enclosed in this fashion, the total area enclosed is the same as

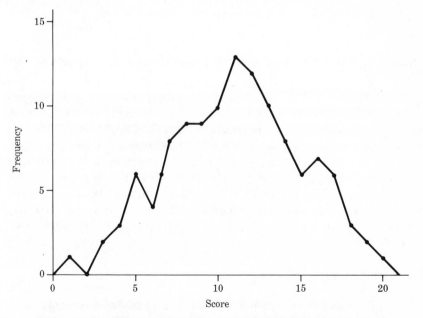

Figure 5.4   Frequency Polygon of 120 English Usage Test Scores.

that of the histogram for the same data. The student should note, however, that the area of the various intervals is not precisely recorded in the
frequency polygon, as it is in the histogram.

Figures 5.4 and 5.5 are frequency polygons constructed from the same
data used to construct the histograms in Figs. 5.1 and 5.2.

**Figure 5.5   Frequency Polygon of 120 Scores Grouped in Intervals of Two
Points.**

The histogram portrays the frequency distribution as though all scores
in each interval were uniformly distributed throughout the interval,
while the frequency polygon portrays the distribution as though all scores
were concentrated at the midpoints of the intervals. The frequency polygon, like the histogram, carries the same information as the frequency
distribution from which it was made. Some investigators prefer the frequency polygon when they wish to compare the form of two different frequency distributions. Perhaps the greatest contribution of the frequency
polygon is that its form suggests that the trait portrayed is continuously
distributed, which is often the case.

5.4   THE OGIVE

An *ogive* resembles a frequency polygon, except that the vertical scale is
recorded in cumulative frequencies, relative cumulative frequencies, or

percentile ranks. There is another important distinction: The plotting points on the ogive are located above the upper real limits of the intervals rather than above the midpoints, as used with the frequency polygon. The upper real limits are used, since the graph is intended to represent the number of measures falling below specified score values. Such a graph, if constructed on a sufficiently large scale, may be used in estimating centiles and percentile ranks. Figure 5.6 is an ogive constructed with the data of Table 3.3, which was also used to construct the histogram of Fig. 5.1 and the frequency polygon of Fig. 5.4.

Figure 5.6   Ogive of 120 English Usage Test Scores.

Referring to Fig. 5.6, the student may locate the 50th percentile rank on the vertical axis (where $f = 60$) and determine that the value of the median is approximately 11.1 on the horizontal axis. Similarly, he may determine that the value of Q_1 is approximately 8.2 and that Q_3 is approximately 13.9 on the score scale. If the ogive is constructed specifically for the purpose of estimating centiles and percentiles ranks, the vertical scale should be divided into 100 equal parts and calibrated in percentile ranks.

5.5   GRAPHS OF CONTINUOUS DISTRIBUTIONS

There is substantial evidence that many of the variables studied by behavioral scientists are continuously distributed even though their measuring instruments yield discrete scores which are only approximations of the supposedly continuous variables. When these measurements are made on relatively small samples drawn from very large populations, it is reasonable to assume that the irregularities in a frequency polygon may be attributed to sampling and measurement errors rather than being truly characteristic of the distribution of the trait in the parent population. The frequency polygon of a truly continuous variable in a very large population would be a smooth curve, such as that in Figs. 5.6, 5.8, 5.9, and 5.10.

The investigator has reproduced the frequency polygon of Fig. 5.4 with a dashed line in Fig. 5.7 and has also smoothed the original frequency polygon, producing a continuous curve which he hopes will approximate the true distribution of the variable in the parent population. Of course, there is a danger of obscuring the true distribution when a frequency polygon is smoothed in this fashion.

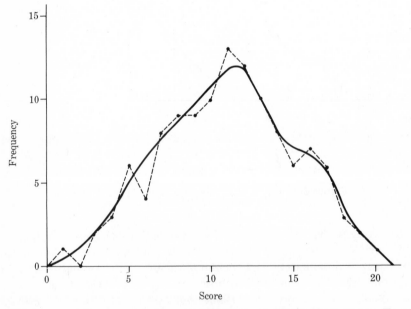

Figure 5.7   Smoothed Frequency Polygon of 120 English Usage Test Scores.

5.6   THE NORMAL CURVE

Certain direct measures used in the behavioral sciences (for example, the height and weight of adult humans) have been demonstrated to closely approximate a mathematical model called the *normal* distribution. The graph of the normal distribution is a continuous, symmetrical, bell-shaped curve. It is high in the middle, indicating a preponderance of frequencies in the vicinity of the median, and low at the ends, indicating low frequencies at both extremes of the distribution. Figure 5.8 illustrates the normal curve.

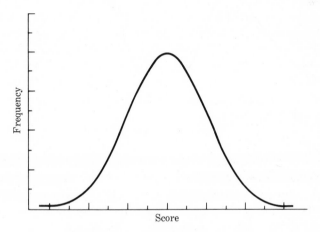

Figure 5.8   The Normal Curve.

The available evidence suggests that many of the traits underlying psychological measures are normally distributed. For this reason, and because the normal distribution is a useful model for statistical analysis, it is common practice among professional test constructors to adjust the difficulty of the items so that standardized tests yield normally distributed scores.

The normal curve is the foundation of statistical theory. It will be treated in considerable detail in subsequent chapters and will prove to be an extremely useful model for both descriptive statistics and statistical inference.

5.7   SKEWNESS AND KURTOSIS

The term *skewness* refers to the symmetry or asymmetry of a curve. If a curve is symmetrical, such as the normal curve, the right and left halves of the curve are mirror images of each other. A curve that is substantially asymmetrical is said to be skewed. If the scores tend to be concentrated toward the low end of the score scale with the curve trailing off to the right, the curve is said to be positively skewed, or skewed to the right. If the scores tend to be concentrated toward the high end of the score scale, the curve is said to be negatively skewed, or skewed to the left. Both types of skewness are illustrated in Fig. 5.9.

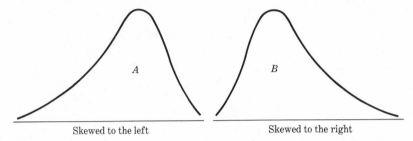

Skewed to the left   Skewed to the right

Figure 5.9   Skewness.

The term *kurtosis* refers to the relative peakedness or flatness of a curve in the center of the distribution. If the curve is flatter than a normal curve, it is said to be platykurtic; if the curve is more peaked than a normal curve, it is said to be leptokurtic. These ideas are graphically illustrated in Fig. 5.10.

Leptokurtic distribution   Normal distribution   Platykurtic distribution

Figure 5.10   Kurtosis.

There are measures of skewness and kurtosis, but these are seldom used by the behavioral scientist and will not be discussed in this text.

EXERCISES

5.1   Refer to the data for Exercise 3.1. Produce a histogram, frequency polygon, smoothed frequency polygon, and ogive from the data. Estimate the values of Q_1, Q_2, and Q_3 from the ogive.

5.2   Refer to the data of Exercise 3.2. Produce a histogram, frequency polygon, smoothed frequency polygon, and ogive from the data. Estimate the values of D_2, D_4, D_6, and D_8 from the ogive.

6

STATISTICAL NOTATION

6.1 THE USE OF SYMBOLS

Symbolic notation makes it possible to state and discuss statistical ideas far more precisely and concisely than is ordinarily possible with common words. The student should realize that the use of symbols is intended to *simplify* the use of statistical procedures. The number of different symbols used is relatively few, and the effort needed to master them will prove to be a profitable investment.

The letter N will be used to represent the number of individuals in a collection of individuals being studied. When the collection is divided into subgroups, N will represent the number in the total collection, while n_1 will represent the number in group one, n_2 the number in group two, etc. If k is the number of groups, then n_k is the number of individuals in the last group. This is a convenient notation for establishing general procedures applicable to the situations where the number of groups may differ. The symbol n_j may be used to represent the number of individuals in *any* group. This general notation permits the establishment of procedures that apply to any or all groups from group one to group k.

The letter X is the symbol most often used to represent the value of a measure in the collection (a test score, for instance). Through the use of subscripts, any score in the collection may be designated. Thus, X_1 represents the score of the first individual, X_2 the score of the second, etc. X_N is the last score in the collection; X_i may be used to represent the score of *any* individual. This general notation permits the establishment of procedures that apply to any or all individuals from the first to the Nth.

When the investigator is confronted with the situation in which there are two measures on each individual, he will probably let Y represent the value of the second measure and use subscripts with it in the same fashion as with X in the previous paragraph.

6.2 THE SUMMATION OPERATOR

Statisticians are so often concerned with summing a collection of measures that a special symbol, the *summation operator*, has been adopted to indicate summation. It is the upper case Greek letter sigma (Σ).

One of the most common uses of the summation operator is to indicate that all of the scores in a distribution are to be added together:

$$\sum_{i=1}^{N} X_i = X_1 + X_2 + X_3 + \cdots + X_N = T$$

The symbols above and below the summation sign indicate the limits of summation, in this case summing all scores from the 1st to the Nth. The idea may be stated in narrative fashion as "taking the sum of the X sub i's as i goes from 1 to N."

When the sum of all measures is to be taken, or when the limits of summation are clearly understood from the context (which is most of the time), statisticians tend to omit the symbols above and below the summation operator and write simply ΣX_i or ΣX. We will also use the upper case letter T (for total) to represent the sum of all scores in a distribution.

The summation operator is used in precisely the same fashion to sum the *number* of individuals in all the subgroups to obtain the total number of individuals:

$$\sum_{j=1}^{k} n_j = n_1 + n_2 + n_3 + \cdots + n_k = N$$

Another common use of the summation operator is to indicate that all of the scores in a collection are to be squared first and then summed:

$$\sum_{i=1}^{N} X_i^2 = X_1^2 + X_2^2 + X_3^2 + \cdots + X_N^2$$

Notice that the operation indicated by the exponent is performed before the operation indicated by the summation symbol.

Still another common use of Σ is when the scores are first summed and then the sum is squared:

$$(\Sigma X_i)^2 = (X_1 + X_2 + X_3 + \cdots + X_N)^2 = T^2$$

Notice the difference between this procedure and the preceding one. Here, the operation indicated by the summation symbol is performed before the operation indicated by the exponent. Of course, we can use the symbol T^2 to indicate the square of the summed scores.

Finally, consider the situation in which two measures are available on each individual, and the statistician wishes to take the *product* of each individual's scores and then sum the products for the entire collection:

$$\Sigma X_i Y_i = X_1 Y_1 + X_2 Y_2 + X_3 Y_3 + \cdots + X_N Y_N$$

If the desired operation calls for summing the scores first and then taking the product, the notation would be as follows:

$$(\Sigma X_i)(\Sigma Y_i) = (X_1 + X_2 + \cdots + X_N)(Y_1 + Y_2 + \cdots + Y_N)$$

This operation could also have been indicated by the product $T_x T_y$, where T_x represents the sum of the X-scores and T_y the sum of the Y-scores.

Table 6.1   The Use of the Summation Operator

i	X_i	Y_i	X_i^2	Y_i^2	$X_i Y_i$
1	3	2	9	4	6
2	0	1	0	1	0
3	4	3	16	9	12
4	2	2	4	4	4
5	3	5	9	25	15
Sums:	$\Sigma X = 12$	$\Sigma Y = 13$	$\Sigma X^2 = 38$	$\Sigma Y^2 = 43$	$\Sigma XY = 37$

The student should study carefully the operations indicated in Table 6.1 so that he understands fully the manner in which these operations were carried out. The scores indicated under the columns X_i and Y_i are the raw data from which the table was constructed.

6.3   RULES FOR THE SUMMATION OPERATOR

Rule 1   The summation of an algebraic sum is the algebraic sum of the summations. This is sometimes referred to as distributing the operator. For example:

$$\Sigma (X_i + Y_i - Z_i) = \Sigma X_i + \Sigma Y_i - \Sigma Z_i$$

Rule 2   The summation of the product of a variable and a constant is the product of the constant and the summation of the variable. For example:

$$\Sigma C X_i = C \Sigma X_i$$ *sum X + the multiply by constant*

Rule 3   The summation of a constant is the constant multiplied by the number of terms in the summation. For example:

$$\Sigma C = NC$$

Let us now illustrate the use of all three rules with a single statistical expression. Given:

$$\Sigma (X_i - M)^2 \qquad \text{where } M \text{ is a constant}$$

First, the operation indicated by the exponent:

$$\Sigma (X_i^2 - 2MX_i + M^2)$$

Next, the distribution of the operator according to Rule 1:

$$\Sigma X_i^2 - \Sigma 2MX_i + \Sigma M^2$$

Then, the application of Rule 2 to the second term of the expression:

$$\Sigma X_i^2 - 2M \Sigma X_i + \Sigma M^2$$

Finally, the application of Rule 3 to the last term:

$$\Sigma X_i^2 - 2M \Sigma X_i + NM^2$$

6.4 DOUBLE SUMMATION

Often the investigator will have a collection of measures that are divided into subgroups—control and experimental groups, for example. Such a collection is subject to double classification (one classification to identify the group, the other to identify the individual in the group) and will probably necessitate the use of double subscripts on some variables and the use of double summation operators. It is generally useful to think of this type of data as organized into rows and columns, with the first subscript indicating the row and the second the column in which the measure is found. Thus, X_{32} represents the score in the third row, second column. If the subgroups are organized into columns (as in Table 6.2), then X_{32} is

Table 6.2 Data Organized in Subgroups

i	Group I	Group II	Group III
1	5	0	2
2	3	1	4
3	2	3	5
4	4	2	3
Totals	14	6	14

the score of the third person in the second group. Most of the data presented in this book will be organized in this fashion, with the first subscript indicating the individual and the second the group. There is no reason, however, why the data could not be organized with the groups in rows, in which case X_{32} becomes the second person in the third group. Whichever way the data are organized, the first subscript will always indicate the row number and the second the column number.

Referring to the data in Table 6.2, we immediately see that n_1 is 4, n_2 is 4, n_3 is 4, and N is 12. Let T_1 be the sum of scores in Group I, T_2 the sum of scores in Group II, and T_3 the sum of scores in Group III. Thus, T_1 is 14, T_2 is 6, and T_3 is 14. Let T be the sum of scores for *all* groups; in this case T is 34. We will use the expression T_j to represent the sum of scores in *any* group.

The sum of scores in any column is given by the expression:

$$T_j = \sum_{i=1}^{n_j} X_{ij}$$

For instance, summing the scores in Group II:

$$T_2 = \sum_{i=1}^{4} X_{i2} = 0 + 1 + 3 + 2 = 6$$

The sum of all the measures in the entire collection is given by the expression:

$$T = \sum_{j=1}^{k} T_j = \sum_{j=1}^{k} \sum_{i=1}^{n_j} X_{ij}$$
$$= (5 + 3 + 2 + 4) + (0 + 1 + 3 + 2) + (2 + 4 + 5 + 3)$$
$$= 14 + 6 + 14 = 34$$

Whenever more than one summation operator is used in this fashion, the inside operator (the one nearest the variables) will be distributed first.

EXERCISES

6.1 Given the collection of scores: 0, 1, 2, 4, 5, 6, 7, 8, 8, 9, evaluate the following expressions:

(a) ΣX_i

(b) ΣX_i^2

(c) $(\Sigma X_i)^2$

(d) $\Sigma X_i^2 - \dfrac{T^2}{N}$

6.2   Given the following collection of scores in which there are two measures on each of six individuals:

i	X_i	Y_i
1	90	1.2
2	97	2.2
3	104	2.3
4	113	3.9
5	118	3.0
6	125	3.5

Evaluate the following expressions:
(a) ΣX_i
(b) ΣY_i
(c) $\Sigma X_i Y_i$
(d) $(\Sigma X_i)(\Sigma Y_i)$
(e) $\Sigma XY - \dfrac{(\Sigma X)(\Sigma Y)}{N}$

6.3   Given the following collection of scores:

i	Group I	Group II	Group III	Group IV
1	2	1	0	2
2	5	1	3	3
3	4	3	2	0
4	4	4	0	1
5	0	5	4	6
6		2	4	0
7		3	3	5
8		1		2
9		0		
10		5		

Evaluate the following expressions:
(a) Σn_j
(b) ΣX_{i3}
(c) $\Sigma\Sigma X_{ij}$
(d) $\Sigma\Sigma X_{ij}{}^2$
(e) $(\Sigma\Sigma X_{ij})^2$

MEASURES OF CENTRAL TENDENCY

7.1 INTRODUCTION

One of the most useful ideas in statistics is the representation of a collection of measures by a measure of *central tendency*, that is, a single value chosen in such a fashion as to be *typical* of the collection. Just what constitutes a typical (or average) measure will depend to a large extent on the level of measurement and the manner in which the measures are dispersed throughout the collection. There are three average measures or measures of central tendency in common use: the *mean*, *median*, and *mode*. The student is already familiar with these, although he may not have called them by these names.

When we say that the "average" IQ of a group of students is 105, it is usually understood that the term average refers to the arithmetic mean, which we will call simply the *mean*. The mean of a collection of scores is obtained by adding all of the scores together, then dividing by the number of scores.

When we talk about the "average" teacher's salary, it is usually more meaningful to give the median salary. The median measure is that which divides the collection of measures into two groups of equal size with half of the measures being larger than the median and the other half being smaller.

Suppose, however, that a group of people were divided into three groups as follows: 20 percent of them being college graduates, 55 percent being high school graduates but not college graduates, and 25 percent having less than a high school education. Here, the "average" person is a high school graduate, and the term average refers to the mode—the category in which more people are found than any other category.

7.2 THE ARITHMETIC MEAN

By far the most useful of the measures of central tendency is the arithmetic mean. As a general rule, when the behavioral scientist uses the term *average* he means the mean. The mean will be represented by either of two symbols: the uppercase letter M, or the variable symbol with a bar across the top, for example $\bar{X}$ (called "X-bar"). The mean of a collection of measures is simply the sum of the measures divided by the number of measures, and it may be calculated by the formula:

$$M = \bar{X} = \frac{\Sigma X}{N} = \frac{T}{N}$$

If the number of measures is large, it may be desirable to arrange the scores in a frequency distribution and calculate the mean by the formula:

$$M = \frac{\Sigma f_i X_i}{N}$$

where X_i is the score of any interval and f_i is the frequency of the same interval.

The mean may be thought of as the score that would be assigned to each individual if the total for the collection were to be evenly divided among all individuals. It is the only one of our three measures of central tendency that is algebraically defined, and this permits its use in a wide range of calculations. Of course, this also implies that the measures from which the mean is calculated are on either an interval or a ratio scale. It is also the only one of the three that is dependent upon the value of every measure in the collection. This latter characteristic is usually a valuable one, but not always, as the mean is much affected by extreme scores. Thus, the mean would not be very typical of a student's work if the teacher makes a practice of awarding zeros when the student is absent or otherwise unproductive. Similarly, it would not be very meaningful to talk about the mean annual income of five school board members if the incomes were $150,000, $6000, $5000, $4500, and $3000. The mean of this collection is $33,700, which is hardly representative of the group.

The mean has other attributes which make it very valuable in statistical inference. Whenever a sample is drawn from a parent population with the intent that a sample statistic provide an estimate of a population parameter, it is highly desirable that the sample be genuinely representative of the population and that the statistic be a good, *unbiased* estimator of the parameter. The usual procedure for insuring that the sample is representative of the population is to draw a *random sample*. This is accomplished by drawing names from a hat or by some more sophisticated procedure. (Random sampling will be discussed in greater detail in a later chapter.) When random samples are drawn, the sample mean provides a good, unbiased estimator of the population mean. A discussion of these characteristics follows.

A *stable* statistic is one that does not vary appreciably from sample to sample when several samples are drawn from the same parent population. This suggests that one may have some confidence that such a statistic is a good estimator of the corresponding parameter when a single sample is drawn. The sample mean is generally a more stable statistic than either the sample median or mode.

The value of some statistics is systematically influenced by the size of the sample. The usual situation is one in which the statistic tends to be smaller with smaller samples, grossly underestimating the parameter with very small samples. A statistic that is a function of sample size is said to be biased. The sample mean is an unbiased statistic; that is, it is just as likely to be larger than the population mean as it is to be smaller. The error involved in estimating a parameter from an unbiased statistic is random, and the probability of such error is readily determined. This topic will be treated in considerable detail in a later chapter.

It is general statistical practice to use Greek letters to represent parameters and Roman letters to represent statistics. When the behavioral scientist wishes to draw a distinction between the population mean and the sample mean, he will use μ (Greek lowercase mu) to represent the population mean and M or $\bar{X}$ to represent the sample mean. Generally, in statistical inference, the value of the parameter is unknown, and the Greek symbol is used in phrasing hypotheses about it. There are times, in descriptive statistics and some measurement applications, when the distinction between sample and population does not exist, and the symbols may be used interchangeably.

7.3 MATHEMATICAL OPERATIONS WITH THE MEAN

Rule 1 The general mean (M) of a collection of scores that has been divided into several subgroups is the sum of the products of each subgroup mean (M_j or $\bar{X}_j$) multiplied by the number of measures (n_j) in that subgroup and then divided by the total number of scores (N) in all subgroups:

$$M = \frac{\Sigma\, n_j M_j}{N} = \frac{\Sigma\, T_j}{N} = \frac{T}{N} \qquad \text{where } n_j M_j = T_j$$

Rule 2   Given a collection of scores that has been divided into several (k) subgroups each having the same number (n) of measures, the mean of the entire collection is the mean of the subgroup means:

$$M = \frac{n \, \Sigma \, M_j}{N} = \frac{\Sigma \, M_j}{k} \qquad \text{where } N = nk$$

Rule 3   When a constant (C) is added to each of a collection of scores, the mean of the new scores ($X + C$) is equal to the mean of the original scores plus the constant ($M_x + C$):

$$M_{x+c} = M_x + C$$

Rule 4   When each of a collection of scores is multiplied by a constant (C), the mean of the new scores (CX) is equal to the mean of the original scores (M) multiplied by the constant (CM):

$$M_{cx} = CM$$

7.4   THE MEDIAN

Statistical textbooks differ as to the definition of the median. The problem derives from the fact that the median is not algebraically defined, and there is seldom true consensus with respect to nonalgebraic definitions. One of the more satisfying definitions is that the median is a point on the score scale below which one-half of the scores fall. This definition makes the median synonymous with the 2d quartile, 5th decile, and 50th centile.

The median is often popularly defined as the middle score in a collection. If there is an odd number of scores, the value of the median is simply chosen equal to the middle score. If there is an even number of scores, the value of the median will usually be chosen as a point halfway between the two middle scores. This is not a very satisfactory approach to the problem from a mathematical viewpoint, and some problems arise when more than one person has the same score.

While the median may be used with interval or ratio measures, it is more often used with ordinal measures. The techniques of statistical inference which call for the use of the median often have an underlying assumption that no two persons may have the same score (this is a characteristic of truly continuous data). When this assumption is met, the value of the median may be determined by the method described in the preceding paragraph.

The determination of the value of the median when many persons have equal scores requires a bit more effort. The data should first be arranged in a frequency distribution with unit intervals. If the cumulative frequency of any interval is $N/2$, the median is the upper real limit of that interval. If the cumulative frequency at the midpoint of any interval is $N/2$ (that is, if the interval score has a percentile rank of 50), the median is the same as the interval score. However, if the data do not match either of these criteria, it is necessary to determine the median interval and calculate the median by means of the formula which follows. The median interval is the one with the smallest cumulative frequency larger than $N/2$.

$$\text{median} = X_u - \frac{(cf_i - .5N)}{f_i}$$

where X_u is the upper real limit, f_i the frequency, and cf_i the cumulative frequency of the median interval. The student should realize that this formula does not constitute an algebraic definition of the median but is only a convenient procedure for assigning a numerical value to an indeterminate measure.

The median will usually be preferred over the mean when the data are on an ordinal scale or when extreme scores render the mean meaningless.

7.5 THE MODE

The mode of a collection of measures is that measure which occurs most often. Although the mode may be used with ordinal, interval or ratio measures, it is the only one of the measures of central tendency that may be used with nominal data.

The mode may be highly descriptive of a collection of data, but it is unlikely to enter into statistical calculations, and its use is likely to be restricted to descriptive statistics.

There are many instances in which no single mode can be identified; there may be two or more modes or no identifiable mode at all.

7.6 CALCULATION OF THE MEAN, MEDIAN, AND MODE

The following calculations illustrate the determination of the mean, median, and mode from a frequency distribution. The data are from Table 7.1.

$$M = \frac{T}{N} = \frac{337}{65} = 5.18$$

$$\text{median} = 5.5 - \frac{35 - 32.5}{12} = 5.29$$

$$\text{mode} = 5 \text{ and } 7 \quad \text{(the distribution is bimodal)}$$

Table 7.1   Frequency Distribution of 65 Scores

X	f	fX	cf
9	2	18	65
8	6	48	63
7	12	84	57
6	10	60	45
5	12	60	35
4	9	36	23
3	7	21	14
2	4	8	7
1	2	2	3
0	1	0	1
	$N = 65$	$T = 337$	

The student should take particular note of the fact that changing the value of a score at either extreme without shifting it to the other side of the median will have no effect on the value of the median; however, it will change the value of the mean.

7.7 SELECTING A MEASURE OF CENTRAL TENDENCY

The first consideration in selecting a measure of central tendency is the level of measurement available. If the data are on a nominal scale, only the mode may be used. If the data are ordinal, the mode or median may be appropriate. With interval data, the investigator is privileged to use the mean, median, or mode. Occasionally, it will be desirable to report two or three different measures of central tendency.

The second consideration in selecting a measure of central tendency is the use to which it will be put. If it is intended to be descriptive, then the measure of central tendency most descriptive of the data should be reported. The important idea here is that the measure of central tendency should be typical of the collection that it represents.

If the investigator wishes to draw inferences from sample to population, his choice of a measure of central tendency will be largely determined by the statistical procedures that are appropriate to his data and hypotheses. The student will have opportunity to acquaint himself with these procedures as he progresses into the study of statistical inference.

The mean has a number of distinct advantages. First, because it is algebraically defined, it lends itself to a variety of mathematical operations. Most of our statistical procedures call for these mathematical operations, and the use of the mean is mandatory. The mean has another quality of no small importance in drawing inferences from samples to populations—the quality of stability. This means that the mean of a sample is more likely to be a good estimate of the corresponding population parameter than the other measures of central tendency. However, the mean may be greatly influenced by the presence of extreme scores and cease to be typical of the collection. This is especially true with small samples, and in such cases the median may be preferred.

Figure 7.1 illustrates the relationship between the mean, median, and mode in symmetrical distributions. Curve *A* is unimodal, and Curve *B* is bimodal. Figure 7.2 illustrates the relationship between the mean, median, and mode in assymetrical distributions. Curve *A* is skewed to the right, and Curve *B* is skewed to the left. Collectively, these curves suggest an alternate definition of the median: it is that point on the score scale such that a line perpendicular to the point will divide the area under the curve into two equal parts.

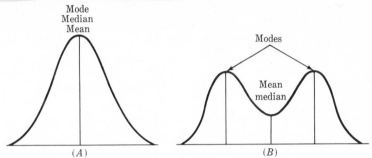

Figure 7.1   Mean, Median, and Mode in Symmetrical Distributions.

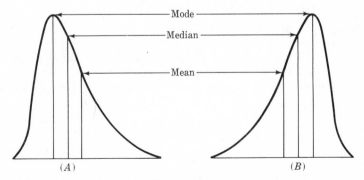

Figure 7.2   Mean, Median, and Mode in Asymmetrical Distributions.

EXERCISES

7.1 Given the following collection of scores: 9, 8, 8, 7, 6, 5, 5, 5, 4, 3,
    (a) Calculate the arithmetic mean.
    (b) Determine the value of the median.
    (c) Determine the mode of the distribution.
    (d) If each score in the distribution were multiplied by 5, what would the value of the mean of the new scores be?
    (e) If 10 were added to each score in the distribution, what would the value of the mean of the new scores be?
    (f) If 6 were subtracted from each score in the distribution, what would the value of the mean of the new scores be?

7.2 Given the following distribution of scores:

X_i	f_i
20	1
18	1
17	3
16	2
14	4
12	5
11	5
10	6
9	4
7	3

    (a) Calculate the arithmetic mean.
    (b) Determine the value of the median.
    (c) Determine the mode of the distribution.

7.3 Refer to the data of Exercise 6.3.
    (a) Calculate the mean for each of the subgroups.
    (b) Using Rule 1 of Section 7.3, determine the general mean of the entire collection.
    (c) Calculate the mean of the entire collection from the formula $M = T/N$.

MEASURES OF DISPERSION

8.1 Introduction

The essential characteristic of any variable is that it varies—that it yields different measures for different individuals. The populations that the behavioral scientist studies will always vary on one or more dimensions, and he will be interested in an index which accurately describes the manner in which they vary. It is customary to refer to this index as a *measure of dispersion*. Often, a fairly adequate summary of a collection of measures may consist simply of a measure of central tendency and a measure of dispersion. These will usually be the mean and an index which describes the dispersion of the measures about the mean.

Six measures of dispersion will be discussed. None of these is appropriate for use with nominal variables. The dispersion of nominal measures is probably adequately treated by reporting the proportion of the measures in each classification or by drawing a bar graph that pictorially expresses the dispersion of the measures. Two of the six measures of dispersion, the range and the semi-interquartile range, are suitable for use with ordinal variables. The other four assume measurement on at least an interval scale.

8.2 THE RANGE

The range is the simplest measure of dispersion, and it is defined as the difference between the largest and the smallest scores in the collection. This may be expressed in algebraic form by $R = X_h - X_1$, where X_h is the highest score and X_1 the lowest. Some statisticians choose to define the range in such a fashion as to include both the highest and lowest scores, in which case it is one larger than the range yielded by our definition. We will be consistent in using the first definition, but the student should be aware that the other definition may appear in the literature.

Because it uses only two scores out of the entire collection, the range may not be very descriptive of the total population. As a general rule, good statistical practice demands the use of as much data as possible. There is also a tendency for the range to vary with sample size—larger samples tending to yield more extreme values. Because the range is totally determined by the two extreme scores, which tend to vary considerably from sample to sample, it is not a very stable statistic. Because the size of the range is a function of sample size, it is said to be a *biased* statistic. These latter characteristics are of importance only when a sample is used to represent a population.

There are times, however, when the range is a highly descriptive measure of dispersion and should be reported. There are also a few nonparametric procedures that use the range and that are used to draw inferences from samples to populations.

8.3   THE SEMI-INTERQUARTILE RANGE

It has been suggested that the range may not be very descriptive of a distribution because its value is totally determined by the two extreme scores. For the same reason it is not a very stable statistic, and it tends to be a function of sample size. These shortcomings are to some extent overcome by selecting the semi-interquartile range as a measure of dispersion.

The term *interquartile range* refers to the interval $Q_3 - Q_1$ which contains half the scores in the distribution. The value of this statistic is not at all influenced by the values of the extreme scores. Dividing the interquartile range by two yields the semi-interquartile range, as indicated by the formula:

$$Q = \frac{Q_3 - Q_1}{2}$$

Consider the two distributions of scores in Table 8.1. The measures are from an attitude scale dealing with race relations. Group I consists of 20 white male union members, and Group II consists of the wives of these men. With a little effort one may quickly determine that the median is 5 and the range is 8 for both groups, which suggests that the two groups do not differ in their attitudes on race relations. With a little more effort one

Table 8.1   Attitudes on Race Relations

Group I (Males)		Group II (Females)	
X_i	f_i	X_i	f_i
9	1	9	1
8	1	8	1
7	3	7	0
6	3	6	3
5	4	5	10
4	3	4	2
3	2	3	2
2	2	2	0
1	1	1	1

may determine that the semi-interquartile range is 1.5 for Group I and 0.5 for Group II, which suggests that the attitudes of the females are not as widely dispersed as those of the males. The behavioral scientist would probably interpret this finding as suggesting that the females are less inclined to choose extreme positions—that they are more homogeneous than the males on the trait being measured.

8.4   THE MEAN DEVIATION

Behavioral scientists make a great deal of use of *deviation scores*—scores calculated as the difference between a raw score and the mean. A deviation score is usually represented by the lowercase form of the symbol used to represent the raw score. The formula for a deviation score is:

$$x_i = X_i - M$$

Persons who score below the mean receive a negative deviation score, and persons who score above the mean receive a positive deviation score. The algebraic sum of all the deviation scores in a collection is equal to zero:

$$\Sigma\, x_i = 0$$

The *mean deviation* is a measure of dispersion which is equal to the mean of the *absolute* values of the deviation scores. An absolute value is one in which the algebraic sign of the variable is disregarded (all values are treated as positive values). The mean deviation is expressed symbolically as follows:

$$MD = \frac{\Sigma |X_i - M|}{N}$$

Absolute values are used, since the mean of the algebraic values is zero. The use of absolute values in statistical calculations is to be avoided, as it precludes the possibility of additional calculations. (This is a mathematical restriction upon the use of absolute values.) For this reason, the mean deviation is used primarily to introduce the measures of dispersion which follow.

8.5   THE SUM OF SQUARES

One way to avoid working with absolute values is to square the deviations before summing. This is common statistical practice, and the student can expect to perform this operation many times. The simplest measure of dispersion using this procedure is the sum of the squared deviations from the mean, which statisticians abbreviate to *sum of squares*.

There are several mathematically equivalent expressions for the sum of squares, and the student must be thoroughly familiar with all of them:

$$SS = \Sigma x_i^2 = \Sigma (X_i - M)^2$$

$$SS = \Sigma X_i^2 - \frac{(\Sigma X_i)^2}{N} = \Sigma X_i^2 - \frac{T^2}{N} = \Sigma X_i^2 - NM^2$$

The several formulas for the sum of squares all get the same results, and the choice of formula is usually determined by the nature of the data and the availability of calculating equipment. Consider the data in Table 8.2. Because these data are simple, the various formulas for the sum of squares are about equally convenient. When an electronic computer or desk calculator is available, it will usually be most convenient to calculate the sum of squares in this fashion:

$$SS = \Sigma X_i^2 - \frac{T^2}{N} = 340 - \frac{2500}{10} = 90$$

Precisely the same result is achieved when we calculate the sum of squares by the formula:

$$SS = \Sigma X_i^2 - NM^2 = 340 - 10(25) = 90$$

Table 8.2   Calculation of the Sum of Squares

i	X_i	X_i^2	x_i	x_i^2
1	0	0	-5	25
2	1	1	-4	16
3	2	4	-3	9
4	4	16	-1	1
5	5	25	0	0
6	6	36	1	1
7	7	49	2	4
8	8	64	3	9
9	8	64	3	9
10	9	81	4	16
$N = 10$	$T = 50$	$\Sigma X^2 = 340$	$\Sigma\ x = 0$	$\Sigma\ x^2 = 90$

Notice in Table 8.2 that it was convenient to calculate $\Sigma\ x_i^2$ directly, but only because the scores and the mean were integers. While it is not at all uncommon for the scores to be integers, it is much more likely that the mean will not be an integer, and this greatly complicates the calculation of the squared deviation scores. However, there is a computational scheme which can be used to simplify the process. For example, consider the data in Table 8.3, which have a mean of 3.2. Let C be an integer constant approximately equal to the mean. In the example, we have let C equal 3. Then, let w_i be a deviation score, the deviation of X_i from C. Thus, $w_i = X_i - 3$. The sum of squares may now be calculated by the formula:

$$SS = \Sigma\ w_i^2 - \frac{(\Sigma\ w_i)^2}{N} = 7 - \frac{1}{5} = 6.8$$

Table 8.3   Calculation of the Sum of Squares

i	X_i	X_i^2	w_i	w_i^2
1	2	4	-1	1
2	2	4	-1	1
3	3	9	0	0
4	4	16	1	1
5	5	25	2	4
$N = 5$	$T = 16$	$\Sigma X^2 = 58$	$\Sigma\ w = 1$	$\Sigma\ w^2 = 7$

That this is in fact the sum of squares may be demonstrated by calculating the sum of squares with one of the other formulas:

$$SS = X_i{}^2 - \frac{T^2}{N} = 58 - \frac{256}{5} = 6.8$$

The computational scheme just described will probably be preferred when the sum of squares must be calculated without the use of computational aids such as a desk calculator or electronic computer.

Although the sum of squares enters into many statistical calculations, including the measures of dispersion which follow, it is not usually reported as a descriptive index. The variance and standard deviation which follow are the most important measures of dispersion in both descriptive and inferential statistics.

8.6 THE VARIANCE

The variance is defined as the mean of the squared deviations from the mean, and this is usually abbreviated to *mean square*. The population variance is represented by σ^2 (σ is the lowercase Greek letter sigma), and it is obtained by dividing the sum of squares by the number of subjects:

$$\sigma^2 = \frac{SS}{N} \quad \text{where } SS = \Sigma \, (X_i - \mu)^2$$

In statistical inference where the sample variance is used to estimate the population variance, it is not sufficient simply to substitute the sample mean for the population mean in the formula above. Such a procedure would result in a biased statistic, one which is systematically influenced by sample size, tending to underestimate the population variance with small sample sizes. In order for the sample variance to be used as an estimate of the population variance, the systematic bias introduced by small sample size must be corrected for by adjusting the formula, that is, substituting $N - 1$ for N. It can be demonstrated both mathematically and empirically that this correction provides an unbiased estimate of the population variance. We will use the symbol S^2 to represent a sample variance which has been calculated in this fashion; thus the formula becomes:

$$S^2 = \frac{SS}{N - 1} \quad \text{where } SS = \Sigma \, (X_i - M)^2$$

The sample variance formula with the adjustment to provide an unbiased estimate of the population variance will always be used in statistical inference. However, there is some disagreement as to which formula should be used in descriptive statistics and those measurement applications where the collection of measures is not necessarily a sample intended to represent some larger population. The author is partial to the use of the population formula, because this is consistent with the definition of variance as the mean of the squared deviations from the mean, and because it is consistent with mathematical procedures to be considered in subsequent chapters. He also recommends the calculation of the sum of squares as a first step to be completed before the calculation of the variance, regardless of which formula is used.

8.7 THE STANDARD DEVIATION

The standard deviation is simply the square root of the variance—the square root of the mean square, or simply the *root mean square*. It will be represented by the symbols σ and s.

The standard deviation has an advantage over the variance as a descriptive measure of dispersion in that it is expressed in the same units as the original measures. For example, modern intelligence tests are commonly scaled with mean of 100 and standard deviation of 15. A person with a score of 115 is one standard deviation above the mean. The variance in this case is 225, which is not a particularly descriptive piece of information. However, both the variance and the standard deviation will be used a great deal in statistical inference.

According to Tchebycheff's theorem, the proportion of the scores in any frequency distribution which are within h standard deviations of the mean is at least $1 - 1/h^2$. From this, one may determine that at least 75 percent of the scores in any collection will lie within the interval $\pm 2\sigma$ from the mean.

The student is advised to become so familiar with the various formulas for the sum of squares, variance, and standard deviation that he recognizes them instantly when they enter into a series of calculations and so that he can reproduce them at will.

8.8 MATHEMATICAL OPERATIONS WITH THE VARIANCE AND STANDARD DEVIATION

Rule 1 When a constant (C) is added to each of a collection of scores, the variance and standard deviation of the new scores $(X + C)$ remain the same as the variance and standard deviation of the orginal scores:

$$\sigma_{x+c} = \sigma_x$$

Rule 2   When each score in a collection is multiplied by a constant (C), the standard deviation of the new scores (CX) is equal to the constant multiplied by the standard deviation of the original scores:

$$\sigma_{cx} = C\sigma_x$$

Rule 3   When each score in a collection is multiplied by a constant (C), the variance of the new scores (CX) is equal to the constant squared (C^2) multiplied by the variance of the original scores:

$$\sigma_{cx}^2 = C^2\sigma_x^2$$

The student is reminded that subtraction is a special case of addition (the addition of a negative number) and that division is a special case of multiplication (multiplication by a reciprocal). The rules above apply to those situations where a constant is subtracted from each score or where each score is divided by a constant.

EXERCISES

8.1   Given the following collection of scores: 0, 1, 2, 4, 5, 6, 7, 8, 8, 9,
    (a)   Calculate the range.
    (b)   Calculate the semi-interquartile range.
    (c)   Calculate the mean deviation.
    (d)   Calculate the sum of squares.
    (e)   Calculate the variance.
    (f)   Calculate the standard deviation.
    (g)   Suppose that 6 were subtracted from each of the scores in the collection, what would the value of the variance of the new scores be?
    (h)   Suppose that each score were multiplied by 2, what would the value of the variance of the new scores be?
    (i)   Suppose that each score were multiplied by 10, what would the value of the standard deviation of the new scores be?

8.2   Given the following collection of scores: 9, 11, 12, 12, 14, 15, 16, 16, 17, 17, 17, 19,
    (a)   Calculate the sum of squares.
    (b)   Calculate the variance.
    (c)   Calculate the standard deviation.

8.3   Refer to the data of Table 3.7.
    (a)   Calculate the sum of squares.
    (b)   Calculate the variance.
    (c)   Calculate the standard deviation.

9

STANDARD SCORES

9.1 THE IDEA OF A STANDARD SCORE

The numerical value of the score a student achieves on an examination is usually highly arbitrary, having meaning only as it is related to the other scores achieved on the test. One way of relating meaning to this numerical score was discussed in a previous chapter: the use of percentile ranks to indicate the placement of the score in the distribution by stating the percentage of scores in the distribution which are smaller. Another approach commonly used calls for indicating the placement of the score with respect to the mean of the distribution. In the latter procedure, the usual alternatives are to state the score in terms of standard deviations above or below the mean, or to change the mean and standard deviation of the collection to standard values and report the scores as values which conveniently communicate the idea of deviation from the mean.

Score distributions which have been altered to have means and standard deviations of standard value are called *standard scores*. The mathematical operation by which a distribution of scores receives a new mean and standard deviation is called a *score tranformation*. The score transformations to be considered in this chapter are *linear* transformations, which means that the shape of the distribution of the standard scores will be precisely the same as that of the original or raw scores. The student should take particular note of this, as there is a common misconception that standard scores are normally distributed. In order for standard scores to be normally distributed, it is necessary that the original score distribution be normal, or that a nonlinear transformation be used to produce a new distribution that is normal. Normally distributed standard scores play a special role in statistical inference.

In constrast to the percentile ranks considered earlier, standard scores are algebraically defined. Percentile ranks are ordinally scaled and may be derived from raw scores which are on ordinal, interval, or ratio scales. Standard scores produced by linear transformations are intervally scaled and may be derived from raw scores which are on interval or ratio scales. Although a score of zero may play a special role in standard scores, it should not be interpreted as an absolute zero of the type encountered with ratio scales.

9.2 STANDARD SCORES WITH ZERO MEAN AND UNIT STANDARD DEVIATION

One of the most useful and certainly the most common linear score transformations is that which produces a score distribution with mean of zero and standard deviation of one. A score reported on such a scale is essentially reported as so many standard deviations above or below the mean.

Consider the distribution of the deviation scores ($x_i = X_i - M$) from a specified collection of raw scores (X_i). From our studies of mathematical operations with the mean, we know that when a constant (in this case, $-M$) is added to each score in a collection, the mean of the new score distribution is equal to the mean of the original distribution plus the constant (in this case, $M - M$), so that the distribution of the deviation scores has a mean of zero. The standard deviation of this distribution is the same as that of the raw scores, since the standard deviation of a distribution is unaffected by adding or subtracting a constant from each score.

If all of the scores in a distribution are multiplied by a constant, the standard deviation of the new distribution will be equal to the original standard deviation multiplied by the constant. Thus, the standard deviation of any distribution may be changed to unity simply by multiplying each score by the reciprocal of the standard deviation ($1/\sigma$). In the case under discussion, the mean of the distribution is unaffected, since a constant multiplied by zero is still zero.

This discussion leads to the mathematical definition of a standard score (z) having mean of zero and standard deviation of one:

$$z_i = \frac{X_i - M}{\sigma} = \frac{x_i}{\sigma}$$

For example, given a score distribution with mean of 100 and standard deviation of 15, the z-score of an individual with a raw score of 120 may be calculated as follows:

$$z = \frac{120 - 100}{15} = \frac{20}{15} = 1.33$$

This is equivalent to saying that the individual scored 1.33 standard deviations above the mean. Consider another example from the same collection, where the individual scored 90 on the original score scale:

$$z = \frac{90 - 100}{15} = \frac{-10}{15} = -0.67$$

This is equivalent to saying that this individual scored 0.67 standard deviations below the mean. Of course, an individual from this group with a raw score of 100 would have a standard score of zero.

9.3   PROPERTIES OF z-SCORE DISTRIBUTIONS

When the mean and standard deviation are calculated for a given collection of raw scores, and all of the raw scores are transformed into z-scores, the resultant z-score distribution has the following properties:

(1)   The mean of the z-score distribution is zero. This is true not only for z-scores but for deviation scores (where $x_i = X_i - M$) as well.

(2)   Raw scores below the mean become negative z-scores, and raw scores above the mean become positive z-scores. This property also applies to deviation scores.

(3)   Both the variance and the standard deviation of a z-score distribution are one.

(4)   The unit for z-scores is the standard deviation. For example, when z equals two, the score is two standard deviations above the mean.

(5)   When z-scores are calculated for random samples, the range of the z-scores is a function of sample size. Typically, large samples tend to range from -3.00 to $+3.00$, but smaller samples tend to have a smaller range.

(6)   The sum of the z-scores is zero. This property also applies to deviation scores.

(7)   The sum of the squares of the z-scores is equal to the number of scores in the collection ($\Sigma z_i^2 = N$). This is true only if the standard deviation is calculated using N in the denominator (see Sections 8.6 and 8.7).

(8)   The shape of the z-score distribution is precisely the same as that of the original score distribution.

9.4   OTHER LINEAR SCORE TRANSFORMATIONS

The investigator often will regard the use of a score distribution with negative score values as unnecessarily awkward. It is a simple matter to convert z-scores to a new distribution having any convenient mean and standard deviation, eliminating the negative scores in the process. It is customary to refer to such scores by the capital letter Z, as in the following formula:

$$Z_i = M' + z_i S'$$

M' is the mean and S' the standard deviation desired in the new distribution.

Suppose the investigator desires a score distribution with mean of 50 and standard deviation of 10, a common choice that yields a score distribution with all positive values, ranging from approximately 20 to 80, while providing for extreme scores ranging from 0 to 100. For example, the individual with a z-score of -0.67 has a Z-score as follows:

$$Z = 50 + 10z = 50 - 6.7 = 43.3$$

Another commonly used standard score distribution has a mean of 500 and standard deviation of 100. For the example quoted above:

$$Z = 500 + 100z = 500 - 67 = 433$$

Modern intelligence tests are commonly scaled in this fashion with a mean of 100 and standard deviation of 15, a vast improvement over the old intelligence quotient (MA/CA) approach to IQ.

9.5   COMBINING STANDARD SCORES

The scores earned by students on examinations are typically recorded as the number of questions answered correctly, the number of words spelled properly, or the number of problems solved. In addition to having different score units, one might expect the difficulty of the test items to vary from examination to examination. For these and other reasons, the raw scores received on social studies, spelling, and arithmetic tests are not directly comparable, and a composite score for a given student covering several such examinations is highly questionable. In fact, if the tests have markedly different standard deviations, a composite of several arithmetic tests might be questioned.

Standard scores of the type under discussion, however, are pure or abstract numbers having no unit. Furthermore, when calculated for the same group of students, standard scores on different examinations have directly comparable means and standard deviations. For this reason, there is usually no objection to the calculation of a composite score using several standard scores.

Often, the teacher or psychometrist will prefer to assign different weights to different scores when making up a composite test score. Suppose that an instructor gave three regular tests and one final examination during a course of study, and that he wished one of the regular tests to contribute only one-fourth as much as the other two to the course grade, and the final examination to count one and one-half times as much. The composite score might be computed in the following fashion:

$$\text{total score} = 0.25z_1 + z_2 + z_3 + 1.5z_4$$

Of course, the instructor might seek an average rather than a total score; in the example above, the total score would be divided by 3.75. This would not produce an average z-score, however, since means vary less than the scores from which they are calculated, and the standard deviation of these average scores would be less than one.

EXERCISES

9.1   Given the following collection of scores: 1, 4, 4, 5, 5, 6, 7, 8,
   (a)   Calculate the mean and standard deviation.
   (b)   Calculate a z-score for each measure in the collection.
   (c)   Transform all of the scores in the collection to produce a new distribution with mean of 50 and standard deviation of 10.

9.2   Given the following collection of scores: 7, 7, 8, 9, 9, 10, 10, 11, 11, 12, 13, 13,
   (a)   Calculate the mean and standard deviation.
   (b)   Calculate a z-score for each measure in the collection.
   (c)   Transform all of the scores in the collection to produce a new distribution with mean of 100 and standard deviation of 15.

9.3   Refer to the data of Table 4.1.
   (a)   Calculate the mean and standard deviation.
   (b)   Calculate a z-score for each measure in the collection.

10

NORMAL
DISTRIBUTIONS

10.1 THE NORMAL CURVE AS A MODEL

Many years ago, learned men developed the *law of errors* from their observations of events such as archery contests. At first, the law of errors was simply a group of generalizations that were found to hold true for many different kinds of phenomena. These generalizations are: (1) small errors occur more often than large errors; (2) very large errors very seldom occur; (3) errors become progressively less numerous as they become larger; (4) negative errors (errors to the left) are about as numerous as positive errors (errors to the right). Over 200 years ago Abraham DeMoivre defined the law of errors mathematically when he derived the equation of the normal curve.

Consider the situation in which a true-false examination with six questions is administered to a group of students who have absolutely no knowledge of the materials on which they are being tested. Suppose also that the examination is very well written and that no clues are given to assist the individual in arriving at a correct answer without knowledge of the material. In other words, the student can only guess at the answers, and the chance of getting a right answer to any question is one in two. If we can assume that the responses are completely random, we can arrive at certain conclusions with respect to the distribution of the scores from this test. First, the average score for the group of students would be three answers right and three answers wrong, and this score would also be the mode. Statisticians would call this the *expected value*. However, not all students would get this score; some would score higher and others lower. Quite a few students would be expected to score four right, and about the same number would get only two right. Somewhat fewer students would be expected to get either five right and one wrong *or* five wrong and one right. And very few students would be expected to get all of the questions right or all wrong. This is a practical application of the law of errors.

We would expect the frequency distribution of these scores to be unimodal and symmetrical, with the concentration of the scores in the vicinity of the mean (the expected value) of the distribution. Actually, the chances of a student getting any particular score on such an examination are readily calculable from basic probability theory, which will be considered in a later chapter. Figure 10.1 shows both the histogram and frequency polygon for the expected distribution of these scores.

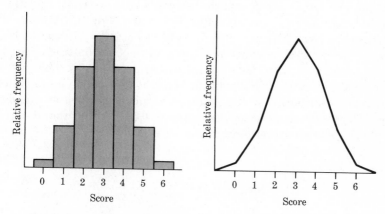

Figure 10.1   Chance Distribution for Six True-False Items.

If the number of questions in the examination were increased to ten, the expected distribution would be as shown in the histogram and frequency polygon of Fig. 10.2. The shape of the frequency polygon suggests

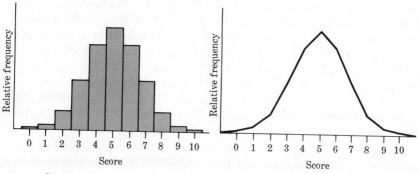

Figure 10.2   Chance Distribution for Ten True-False Items.

that the distribution of the trait being measured is continuous, and that if the polygon were smoothed it would closely approximate a normal curve. As a matter of fact, it can be demonstrated mathematically and empirically that the distribution of these scores would approach a normal distribution if the number of questions and the number of students were both very large.

The normal curve is essentially a mathematical model for chance errors. The frequency distribution of many events found in nature closely approximate the normal distribution, and it has proved to be an extremely useful model for the study of these events. The distribution of scores yielded by psychological tests often closely approximates the normal distribution, and the available evidence suggests that the traits underlying these measures are in fact normally distributed. For this reason, and because the normal distribution is a useful model for statistical analysis, there is often a conscious effect on the part of the test constructor to produce an instrument that yields normally distributed scores. This may be accomplished by adjusting the difficulty of the items or by using a non-linear score distribution which produces normally distributed standard scores. The task is made somewhat easier by the fact that the tests tend to yield scores with rather large chance error components.

10.2   THE EQUATION OF THE NORMAL CURVE

The study of relationships between variables is central to any scientific endeavor, and every scientific equation is an expression of such a relationship. Whenever two variables are related in such a fashion that the value of one variable is determined by the value of the other, the two are said to be *functions* of each other, and an equation can be written to express this relationship. Such a relationship is indicated by the general equation $Y = f(X)$, which says that Y is a function of X. Whenever the values of two variables are paired in this fashion, a graph of the relationship may be drawn. The X-variable will be plotted along the horizontal axis, and the Y-variable along the vertical axis, with a single point on the graph representing each pair of scores. When all the points are connected, a curve is produced which is a graphic representation of the equation. Mathematicians call this curve the *locus* of the equation. Figure 10.3 is a graph of the equation $Y = 2X - 3$. Notice that negative values are as readily plotted on the graph as positive ones. The equation in this case is a simple linear equation, and its locus is a straight line. This is characteristic of equations with two variables, neither of which has an exponent.

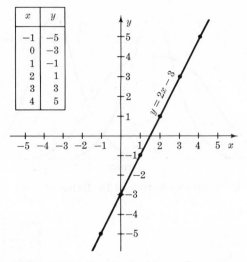

x	y
−1	−5
0	−3
1	−1
2	1
3	3
4	5

Figure 10.3   Graph of a Linear Equation.

The normal curve is a graphic plot of a rather complicated mathematical function. This equation defines a family of curves (infinite in number), any one of which may be completely determined by specifying the values of the mean and standard deviation:

$$Y = \frac{1}{\sigma \sqrt{2\pi}} e^{-(X-\mu)^2/2\sigma^2}$$

where Y = height of the curve corresponding to a given score

X = a score value corresponding to a given height

μ = the mean of the X-variable

σ = the standard deviation of the X-variable

π = a constant approximately equal to 3.1416

e = a constant approximately equal to 2.7183

The role of the mean and standard deviation in determining the particular normal curve is illustrated in Fig. 10.4. Curves A and B have the same mean but different standard deviations. Curves B and C have the same standard deviation but different means.

The equation of the normal curve may be somewhat simplified by specifying that the mean be zero and the standard deviation unity:

$$Y = \frac{1}{\sqrt{2\pi}} e^{-z^2/2}$$

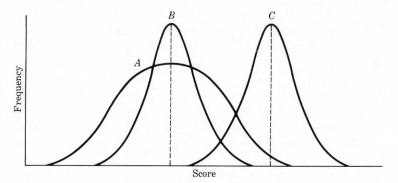

Figure 10.4   Mean and Standard Deviation of Normal Curves.

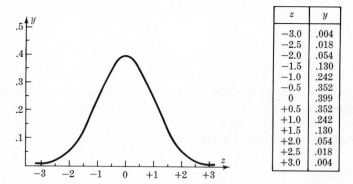

z	y
-3.0	.004
-2.5	.018
-2.0	.054
-1.5	.130
-1.0	.242
-0.5	.352
0	.399
$+0.5$	.352
$+1.0$	.242
$+1.5$	.130
$+2.0$	.054
$+2.5$	.018
$+3.0$	.004

Figure 10.5   Ordinates of the Standard Normal Curve.

This equation in standard score form is sometimes referred to as the *standard normal curve*. Figure 10.5 is a graph of the standard normal curve.

10.3   SUMMARY OF THE PROPERTIES OF NORMAL CURVES

(1)   A normal curve is a graph of a particular mathematical function, a model for events whose outcome is left to chance.

(2)   There is an infinite number of different normal curves, each one determined by the values of the mean and standard deviation. The standard normal curve has a mean of zero and standard deviation of unity.

(3)   A normal curve is symmetrical and bell-shaped with its maximum height at the mean.

(4) A normal curve is continuous; there is a value of Y for every value of X, including fractional values. Its use as a model for frequency distributions assumes the X-variable is continuously distributed.

(5) The value of Y is positive for all values of X. While scores may be positive or negative, there is no such thing as a negative frequency.

(6) The curve approaches but never touches the X-axis (mathematicians say it is *asymptotic* to the X-axis). The score scale ranges from minus infinity to plus infinity; however, the frequency of scores beyond ± 3 standard deviations is near zero.

(7) The inflection points of the curve occur one standard deviation on either side of the mean. Inflection points are points at which the bend in the curve changes direction.

(8) The total area between the curve and score scale is unity, and areas under portions of the curve may be treated as relative frequencies.

10.4 AREAS UNDER THE NORMAL CURVE

In Chapter 5, "Graphs of Frequency Distributions," it was established that the frequencies of the various scores are represented by the areas of the vertical bars in a histogram. This also holds true for graphs of continuous distributions if vertical lines are drawn from the real limits on the score scale to the curve. Of course, the determination of the area of a bar having a curved top is not quite as simple as the determination of the area of a simple rectangle, as encountered in the histogram. Ordinarily, the calculation of the area underneath a portion of a curve requires the use of integral calculus. In the case of the normal curve, this is complicated by the fact that the curve has no simple integral, and the determination of the frequency of a given score or range of scores would be a laborious process even for a mathematician. Fortunately, we are spared the task of these involved calculations, since the areas under the normal curve have been extensively tabled.

Figure 10.6 illustrates the relationship between scores (expressed as standard deviations from the mean) and relative frequencies (expressed as proportions of the area under the curve). Notice that the proportion of the area between the mean and any point on the z-score scale is the same for all normal distributions. For example, 34.13 percent of the scores in a normal distribution fall between the mean and minus one standard deviation. With a little effort, one may readily ascertain that 15.87 percent of the area under the curve is to the *left* of a score of minus one. This means that approximately 16 percent of the normally distributed scores are below a score of minus one standard deviation, and such a

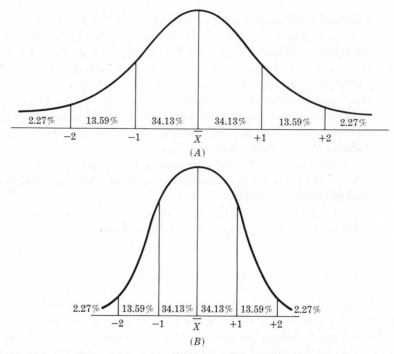

| 2.27% | 13.59% | 34.13% | 34.13% | 13.59% | 2.27% |

$$-2 \qquad -1 \qquad \overline{X} \qquad +1 \qquad +2$$

(A)

2.27%   13.59%   34.13%   34.13%   13.59%   2.27%

$$-2 \qquad -1 \qquad \overline{X} \qquad +1 \qquad +2$$

(B)

Figure 10.6   Areas Under the Normal Curve.

score corresponds roughly to the 16th centile. Similarly, one may determine that a score of plus two standard deviations roughly corresponds to the 98th centile.

Table 3 in the Appendix gives the proportion of the total area under a normal curve between the mean and any score expressed in standard deviations from the mean. Of course, any score expressed in standard deviations from the mean is a z-score; in this case, we are concerned only with the z-scores that are known to be normally distributed. Because the normal curve is symmetrical, the areas from the mean are the same for both positive and negative z-scores, and only positive scores are recorded in the table. To illustrate the use of Table 3, the student should verify that 44.06 percent of the area under the curve is between the mean and $z = 1.56$ and that 49.87 percent of the area is between the mean and $z = -3.02$. The areas are recorded as proportions in the table and must be multiplied by 100 to be reported as percentages.

Now, let us illustrate the use of the table for translating any normally distributed z-score into a percentile rank. If the score is positive, one need only add .5000 (the area to the left of the mean) to the tabled value and multiply by 100 to get the percentile rank. For example, a z-score of .05 corresponds to a percentile rank of 52, a z-score of 1.13 to a percentile rank of 87, and a z-score of 2.06 to a percentile rank of 98. The author has taken the liberty of rounding the percentile ranks to whole numbers. If the score is negative, one need only subtract the tabled value from .5000 and multiply by 100 to get the percentile rank. For example, a z-score of $-.88$ corresponds to a percentile rank of 19 in any normally distributed collection of scores. The important idea is that percentile rank is represented by the area to the *left* of a score. The student may find it easier to understand this idea if he makes a freehand sketch of a normal curve for each of the examples above, carefully locates the mean and the given score on the score scale, draws vertical lines at these two points, which divide the area under the curve into three parts, and uses Table 3 to determine the proportion of the area in each of the three parts. He is reminded that .5000 of the total area is on either side of the mean. Finally, he must determine what proportion of the total area is to the left of the score.

Suppose the investigator wants to determine what proportion of the scores falls between plus one or minus one standard deviation of the mean. To do this, he need only add the area between the mean and plus one standard deviation to the area between the mean and minus one standard deviation. Thus, it may be readily determined that 68.26 percent of the scores fall within ± 1.00, 95.44 percent within ± 2.00, and 99.74 percent within ± 3.00 standard deviations of the mean.

Often, the researcher will want to determine what proportion of the scores falls in the extreme area at one end of the curve or in the extreme areas at both ends of the curve. For example, 2.28 percent of the scores are above $+2.00$, 2.28 percent are below -2.00, and 4.56 percent are beyond ± 2.00 standard deviations of the mean. Again, if the student experiences any difficulty in determining where these values come from, he is urged to make a freehand sketch of a normal curve for each example, divide the area into parts by drawing vertical lines upward from the mean and from the given score or scores, and determine the area in each part. Many investigators find this is a useful technique when working with areas under the normal curve.

EXERCISES

10.1   Given a score point located 1.58 standard deviations above the mean in a normal distribution, determine the following:

(a)   The area between the mean and the score.   .4429

(b)   The total area to the left of the score.   .5000 + .4429 = .9429

(c)   The total area to the right of the score.   .5000 − .4429 = .0571

(d)   The percentile rank of the score.   94 %

10.2   Given a score point located 1.58 standard deviations below the mean in a
normal distribution, determine the following:
(a)   The area between the mean and the score.   . 4429
(b)   The total area to the left of the score.   , 0 571
(c)   The total area to the right of the score.   , 9429
(d)   The percentile rank of the score.   6 %

10.3   Given a score point located 2.81 standard deviations above the mean in a
normal distribution, determine the following:
(a)   The area between the mean and the score.   , 4975
(b)   The total area to the left of the score.   , 9975
(c)   The total area to the right of the score.   , 0025
(d)   The percentile rank of the score.   99%

10.4   Given a score point located 2.81 standard deviations below the mean in a
normal distribution, determine the following:
(a)   The area between the mean and the score.   , 4975
(b)   The total area to the left of the score.   , 0025
(c)   The total area to the right of the score.   , 9975
(d)   The percentile rank of the score.   0 %

10.5   Given a percentile rank of 22 in a normal distribution, determine the
corresponding z-score.   , 77

10.6   Given a percentile rank of 91 in a normal distribution, determine the
corresponding z-score.   1 , 34

10.7   Given a percentile rank of 50 in a normal distribution, determine the
corresponding z-score.   0

10.8   What proportion of the scores in a normal distribution is higher than
$+3.00$? What proportion is lower?   a, , 9987   b, , 0013

10.9   Suppose an investigator had a z-score on each person in a group known to
be normally distributed and that he wished to identify any person in the
top 5 percent of the score distribution. What z-score would be needed to
place a person in the top 5 percent?   z of , 4495 = 1, 64

10.10   Suppose an investigator had a z-score on each person in a group known to
be normally distributed, and that he wished to identify any person in the
middle 50 percent of the group. What z-score range would include the
middle 50 percent of the distribution?   ± , 67

NORMALIZED STANDARD SCORE DISTRIBUTIONS

11.1 INTRODUCTION

Two kinds of score transformation have been discussed in previous portions (Chaps. 4 and 9) of this text. The first, transformation of raw scores
to percentile ranks, was a nonlinear score transformation that altered the
distribution of the scores. No matter what form the distribution of the
raw scores might have, the distribution of the percentile ranks was always
rectangular. The second, transformation of raw scores to z-scores or
Z-scores, was a linear transformation that did not alter the form of the
distribution of the scores. No matter what form the distribution of the raw
scores might have, the distribution of the standard scores was the same.
In this chapter, nonlinear score transformations which produce normally
distributed scores will be considered.

There are several reasons why a behavioral scientist might desire to
transform a collection of raw scores from whatever distribution form they
might have to a normal distribution. He might be convinced that the
underlying trait which is being assessed indirectly by a psychological
instrument is in reality normally distributed and more accurately represented by normally distributed scores. Or, he might desire to make comparisons of measures on several different dimensions of the same population, and this is more readily justified if the measures are all distributed
in the same fashion. Finally, most of the techniques of statistical inference
are specifically intended for use with normally distributed measures, and
the use of these techniques is more readily defended if the score distribution is normal.

There are two commonly used procedures for achieving normally distributed measures. The test constructor may seek to build an instrument that yields normally distributed raw scores, and this is accomplished by manipulating the difficulty of the items. If the underlying trait is indeed normal, this should not be too difficult to do; however, it may require considerable research and revision. As an alternative, the investigator may choose to use a nonlinear transformation that yields normally distributed standard scores, a procedure that usually entails the determination of percentile ranks as an intermediate step.

Several different normal standard score systems are used by behavioral scientists, and the relationships between the most popular ones are illustrated in Fig. 11.1.

Figure 11.1 Popular Systems of Standard Scores.*

11.2 NORMALIZED z-SCORES

To convert raw scores of any form to normalized z-scores, it is first necessary to transform the raw scores to percentile ranks. Once this is

* Figure 11.1 is reprinted from *Test Service Bulletin*, 1955, **48,** The Psychological Corporation, New York.

accomplished, the investigator may refer to a table of areas under the normal curve and translate the percentile ranks into normal z-scores.

For example, suppose a percentile rank of 20 is to be transformed into the equivalent normal z-score. A percentile rank of 20 has 20 percent of the distribution below or to the left of it and 30 percent of the distribution between the score and the mean. Turning to Table 3 in the Appendix, we find that a z-score of $\pm.84$ has .2995 of the total area under the curve between the score and the mean. The normal z-score equivalent to a percentile rank of 20 must be $-.84$, since percentile ranks less than 50 correspond to negative z-scores. For a second example, let us transform a percentile rank of 60 to the equivalent normal z-score. This will be a positive z-score with .1000 of the area under the curve between the score and the mean. Turning to Table 3 again, we find that .25 is the appropriate z-score.

As with a linear transformation, the mean of the normalized z-score is zero, and the standard deviation is unity.

11.3 T-SCORES

T-scores are normally distributed scores with mean of 50 and standard deviation of 10. They may be calculated from normally distributed z-scores by the linear equation:

$$T = 50 + 10z$$

T-scores are especially convenient to use, since all score values are positive, and the range of the scores is usually about 20 to 80. T-scores are customarily rounded to integer values and should be carefully identified to avoid their being mistaken for percentile ranks.

Several similar systems are used with standardized tests. The College Entrance Examination Board, for example, uses a mean of 500 and standard deviation of 100. Deviation IQ scores are calculated using a mean of 100 and standard deviation of 15.

11.4 STANINES

Stanines are normally distributed integer scores with mean of 5, standard deviation of 2, and range from 1 to 9. It is probably most appropriate to think of each stanine as covering a range of scores; a stanine of 5, for example, corresponds to all z-scores in the range $-.25$ to $+.25$. A stanine of 6 corresponds to z-scores in the range $+.25$ to .75, and a stanine of 7 to

z-scores of $+.75$ to $+1.25$. Other values may be easily determined by referring to Fig. 11.1. Stanines are a very crude form of normally distributed scores, but they are especially convenient when large masses of data must be stored on computer punchcards. Each variable recorded in stanine form requires only one column on a punchcard.

EXERCISES

11.1 Refer to the data of Table 4.1. Produce a normalized z-score distribution for this collection of measures.

11.2 Refer to the data of Exercise 3.1.
    (a) Produce normalized z-scores for the entire distribution.
    (b) Produce T-scores for the entire distribution.
    (c) Produce stanines for the entire distribution.

12

PEARSON PRODUCT MOMENT CORRELATION

12.1 THE STUDY OF RELATED VARIABLES

The author likes to think of scientific research as the systematic study
of relationships between variables. The first eleven chapters of this book
have been concerned primarily with the analysis of a single variable.
We have occupied ourselves with the fundamental characteristics of the
variable, some techniques for facilitating its manipulation, and some
procedures for examining its distribution in the group under study.
We now progress to the situation in which there are two or more observations available on each member of the group; here we are particularly
concerned with the possible relationships between these variables. For
example, it is commonly accepted that students with high IQ scores tend
to earn higher grades in school than students with low IQ's. We are
interested in determining whether or not such relationships exist, the
degree of relationship, and the practical implications of any such relationship. A prerequisite to any study of this type is that an observation
on one variable be paired with an observation on a second variable for
each individual in the group under study.

The term *correlation* refers to the degree of correspondence or relationship between two variables. Correlated variables are those which
tend to vary together—when one is larger, the other tends to be systematically larger or smaller. Consider the situation in which an achievement
test is administered to a group of students at the beginning of the school
year and again at the end of the year. Generally, there will be a tendency
for students who scored above the group average at the beginning also
to score above the average at the end. Similarly, most of the students

who scored below the average at the beginning will probably score below the average at the end. The two measures are said to be *positively correlated*. Occasionally, the behavioral scientist will encounter a situation in which a high score on one variable tends to be paired with a low score on another variable. These two variables are said to be *negatively correlated*. Often, whether a correlation is positive or negative is a completely arbitrary thing. A variable that correlates positively with extroversion correlates negatively with introversion. But extroversion and introversion are ordinarily regarded as two different ways of looking at the same variable. The important idea is that there is *concomitant variation* of two correlated variables.

A *correlation coefficient* is an index of relationship between two variables. There are several different kinds of correlation coefficient but most of these have certain characteristics in common. When two variables are perfectly positively correlated, a direct relationship exists such that a higher score on one variable is always associated with a higher score on the other variable. This perfect positive correlation will have a coefficient of one. When two variables are perfectly negatively correlated, an inverse relationship exists such that a higher score on one variable is always associated with a lower score on the other. This perfect negative correlation will have a coefficient of minus one. When there is no relationship between the two variables i.e., when there is no tendency for a higher score on one to be associated with either a higher or lower score on the other, the variables are said to be uncorrelated, and the coefficient is zero. While perfect correlations and zero correlations are sometimes encountered by the physical scientist, most of the variables studied by the behavioral scientist will be correlated, but imperfectly, yielding a correlation coefficient somewhere between zero and plus or minus one.

This chapter is concerned with the most commonly used measure of correlation, the Pearson product moment correlation coefficient, which was named for its originator, pioneer behavioral scientist Karl Pearson. Several other correlation coefficients, some of which are adaptations of the Pearson, will be considered in Chap. 13.

12.2   GRAPHING THE RELATIONSHIP BETWEEN TWO VARIABLES

When two measures are available on each of a group of individuals, the first measure will be designated as the X-variable and the second as the Y-variable. As before, the relationship between the two variables will be graphed with the X-variable plotted along the horizontal axis and the Y-variable along the vertical axis. A single point will be located on the graph for each pair of scores. Figures 12.1–12.5 are graphs constructed in this fashion. A complete listing of the scores from which the graphs were constructed is included with each graph.

Figure 12.1 is a graph illustrating perfect positive correlation for six pairs of scores, and Fig. 12.2 is a similar graph for perfect negative correlation. Figure 12.3 illustrates positive—but less than perfect—correlation, and Fig. 12.4 negative—but less than perfect—correlation.

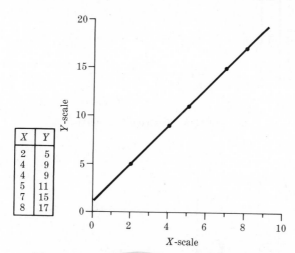

X	Y
2	5
4	9
4	9
5	11
7	15
8	17

Figure 12.1   Perfect Positive Correlation.

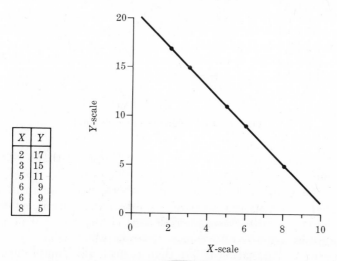

X	Y
2	17
3	15
5	11
6	9
6	9
8	5

Figure 12.2   Perfect Negative Correlation.

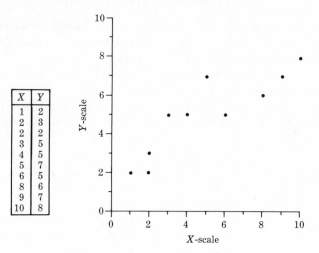

X	Y
1	2
2	3
2	2
3	5
4	5
5	7
6	5
8	6
9	7
10	8

Figure 12.3   Positive Correlation.

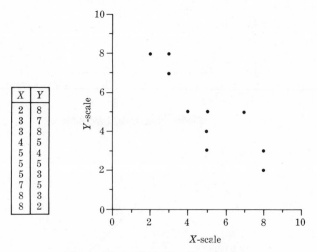

X	Y
2	8
3	7
3	8
4	5
5	4
5	5
5	3
7	5
8	3
8	2

Figure 12.4   Negative Correlation.

Figure 12.5 illustrates the case of two variables with no discernible correlation.

One of the assumptions underlying Pearson product moment correlation is that the relationship between the two variables is a *linear* one. This means that with perfect correlation, either negative or positive, the points on the graph will all lie on a straight line, as in Figs. 12.1 and 12.2. When the correlation is less than perfect, the points depart from the straight-line configuration, as in Figs. 12.3, 12.4, and 12.5. The more

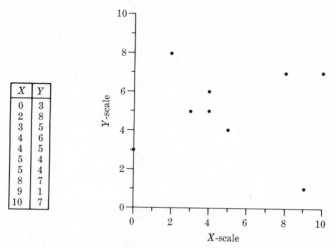

Figure 12.5   Two Uncorrelated Variables.

nearly the pattern of the points approximates a straight line, the higher
the degree of correlation between the two variables. A graph constructed
in this fashion is called a *scatter diagram,* or simply *scattergram.* Just as a
histogram or frequency polygon is a graphic illustration of the manner
in which a single variable varies, a scattergram is a graphic illustration of
the manner in which two variables vary together.

12.3   CALCULATION OF THE PEARSON
CORRELATION COEFFICIENT

By far the most popular of the several available coefficients of correlation is the Pearson correlation coefficient, which may be defined as the
mean of the z-score products of two paired variables. It is represented by
the lowercase letter r and may be calculated from the formula:

$$r = \frac{\Sigma \, z_x z_y}{N}$$

The student is reminded that the sum of the squares of the z-scores in a
given distribution is equal to N (from Section 9.3). Thus, if in a given
collection of paired scores every z_x is equal to the corresponding z_y, the
scores are perfectly positively correlated, and the sum of the z-score
products is the same as the sum of the squared z-scores for either variable.
When the sum of the z-score products equals N (as it would in this case),

the correlation coefficient is unity. Similarly, if every z_x is paired with a z_y of equal magnitude but opposite in sign (every positive score is paired with a negative score), the sum of products will be equal to minus N, and the correlation coefficient will be equal to minus unity.

Although the formula in z-score form serves to illustrate the definition of the correlation coefficient, it will usually be more convenient to calculate the coefficient from this mathematically equivalent formula:

$$r = \frac{SP}{\sqrt{SS_x SS_y}}$$

where

$$SP = \Sigma\, xy = \Sigma\, XY - \frac{(\Sigma\, X)(\Sigma\, Y)}{N}$$

$$SS_x = \Sigma\, x^2 = \Sigma\, X^2 - \frac{(\Sigma\, X)^2}{N}$$

$$SS_y = \Sigma\, y^2 = \Sigma\, Y^2 - \frac{(\Sigma\, Y)^2}{N}$$

The symbols SP and $\Sigma\, xy$ are used to represent the sum of products of the deviation scores, which we will abbreviate to *sum of products*. Notice the close resemblance of the sum of products to the sum of squares, the distinction between the two being the substitution of a product of two variables for the square of a single variable (which is the product of a variable with itself). There is another distinction worthy of note: Although the sum of squares is always positive, the sum of products may be either negative or positive.

Occasionally, it will be convenient to calculate the correlation coefficient from other formulas which are mathematically equivalent:

$$r = \frac{\Sigma\, xy}{N\sigma_x \sigma_y} \quad \text{or} \quad r = \frac{\Sigma\, xy}{(N-1)s_x s_y}$$

The calculation of the sum of products will be illustrated with the data of Figs. 12.1 and 12.2. For Fig. 12.1:

$$SP = 378 - \frac{(30)(66)}{6} = 48$$

For Fig. 12.2:

$$SP = 282 - \frac{(30)(66)}{6} = -48$$

The student will probably profit from completing these calculations for himself from the raw data given with the two graphs.

The calculation of the correlation coefficient will be illustrated with the data of Figs. 12.1, 12.2, 12.3, and 12.4. For Fig. 12.1:

$$r = \frac{48}{\sqrt{(24)(96)}} = 1.00$$

For Fig. 12.2:

$$r = \frac{-48}{\sqrt{(24)(96)}} = -1.00$$

For Fig. 12.3:

$$r = \frac{53}{\sqrt{(90)(40)}} = .883$$

For Fig. 12.4:

$$r = \frac{-34}{\sqrt{(40)(40)}} = -.85$$

The student should complete one or two of these calculations to familiarize himself with the procedure. That the sum of products is zero (a necessary condition for the correlation to be zero) for the data of Fig. 12.5 is an exercise which is left to the student.

12.4 FACTORS WHICH INFLUENCE THE VALUE OF THE COEFFICIENT

The magnitude of the correlation coefficient is unaffected by adding a constant to either or both of the two variables. In fact, some computation effort may be saved by subtracting a constant from each score in one or both of the distributions when the scores have large numerical values.

The magnitude of the correlation coefficient is a function of the variability of the measures. A correlation coefficient calculated from a group having a wide range of talent will be larger than that from a group which is quite homogeneous on one or both of the variables. For example, the correlation between IQ score and an achievement score may be quite high for a group of students ranging in ability from borderline mental retardates to the intellectually gifted. The correlation between the same two variables for a group consisting of only exceptionally bright students is likely to be very low. This suggests that reporting the correlation of two variables is meaningful only when the group under study is specified.

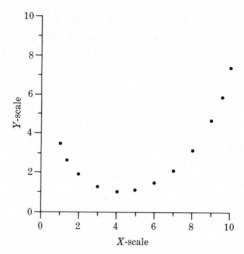

Figure 12.6   Curvilinear Correlation.

The correlation coefficient assumes that the relationship between the two variables is a linear one. Consider the scattergram of Fig. 12.6. There is obviously perfect correspondence between the two variables, but it is not a linear one, and the degree of relationship between the two variables will be grossly underestimated by the calculation of a Pearson correlation coefficient. Curvilinear correlation will be discussed in Chapter 39. As a general rule, whenever the relationship between two variables is sufficiently nonlinear as to invalidate the use of the Pearson correlation coefficient, it will be readily apparent from an inspection of the scattergram.

12.5   INTERPRETATION OF THE PEARSON CORRELATION COEFFICIENT

The student is cautioned against making any absolute interpretation of the correlation coefficient. It is an abstract number, a convenient index of relationship which has been defined in such a fashion as to lie between the limits minus one to plus one. It is invariably expressed as a decimal fraction, but this fraction should not be interpreted as a proportion or percent, and it should not be interpreted in terms of the original score units, since it is independent of the unit and the magnitude of the scores. In fact, there is no need for the two variables to be expressed in the same score units.

Occasionally, one reads that correlation coefficients on the order of .30 to .70 suggest a moderate relationship, while coefficients less than these indicate a low relationship and coefficients larger than these a high relationship. This is a fallacious argument in statistical inference, since the significance of a correlation coefficient is a function of sample size; large correlations achieved with small samples may be completely meaningless. It is also a fallacious argument in descriptive statistics, since the nature of the group and the variables being studied plus the use to which the coefficient is to be put will determine whether a particular coefficient is large or small. For example, a correlation of .70 between scholastic aptitude as measured in the first grade and grade point average in college would be phenomenal. However, a coefficient of .70 between two supposedly equivalent forms of an achievement test would be so low as to suggest major revision is in order. It is important to note also that the degree of relationship is independent of the sign of the correlation coefficient; a coefficient of $-.70$ suggests the same degree of relationship as a coefficient of $+.70$.

Another problem arises in the interpretation of the correlation coefficient because one may add a constant to each score of one of the variables without changing the value of the coefficient. Suppose the investigator wants to determine the relationship between scores from the same test administered to the same group of subjects on two different occasions. A high correlation coefficient might lead the investigator to believe that the two sets of scores are essentially equivalent when, in reality, the scores of the group are significantly higher or lower on the second administration. Similarly, a high correlation between the scores for a group of school children on tests of reading ability and numerical ability does not indicate that the two traits are about equally developed. The correlation coefficient is an index of the concomitant variation of two variables, but it says nothing about the magnitude of the variables.

Possibly the most valid and useful interpretation of the correlation coefficient is achieved by squaring it. With a little effort (and a knowledge of regression), one may demonstrate that r^2 is the ratio of two variances. The denominator of this ratio is the total variance of one of the variables, and the numerator is that part of this variance which can be predicted from or attributed to the other variable. Thus, r^2 is that proportion of the variance of one measure which is accounted for by the other. Consider the situation in which a pretest is administered to a group of students prior to a learning experience, and a posttest is administered to the same group following the learning experience. If a correlation coefficient is calculated from the two variables, r^2 may be interpreted as that proportion of the posttest variance which is accounted for by the pretest. This proportion of the variance of the posttest existed prior to the learning experience and cannot be attributed to the learning which took place between the two test administrations. Suppose that the correlation between an intelligence test score and an achievement test score is .50; then, one might conclude that 25 percent of the variance of the achievement scores is accounted for by intelligence, as measured by the intelligence test. The quantity r^2 is sometimes called the *coefficient of determination*, because it is that proportion of the variance of a variable which may be predicted by or determined by another variable. The quantity $1 - r^2$ is sometimes called the *coefficient of nondetermination*, because it is that proportion of the variance which cannot be predicted by or determined by the other variable.

Possibly the most serious mistake that an investigator can make in the interpretation of a correlation coefficient is to regard a significant correlation as proof of a cause-and-effect relationship. For example, one might investigate the study habits of college students and determine that there is a negative correlation between the amount of time spent in study and course grades. This does not establish that higher grades result from less study, but suggests that brighter students do not study as much as the dull ones. However, the correlation of IQ scores of parents and their children has been reported to be on the order of .50, and the investigator is probably entitled to interpret this as evidence of a causal relationship. The observed correlation between two variables is sometimes due to a cause-and-effect relationship, and this is science at its best; but a significant correlation is not of itself sufficient evidence to establish a causal relationship.

12.6   THE CORRELATION MATRIX

Very often in behavioral research, measures on several variables will be available on each subject in the group being studied, and the investigator will be interested in reporting the correlation between each pair of variables. The most convenient form for reporting a large number of correlations for the same group of subjects is the correlation matrix. A matrix is a rectangular array of numbers arranged in rows and columns, the rows being numbered from top to bottom and the columns from left to right. In the case of the correlation matrix, the rows and columns represent the variables, and the numbers entered in the matrix are the correlation coefficients. Thus, the number entered in the second row and third column is the coefficient of correlation between the second and third variables. Table 12.1 is a correlation matrix reporting all possible correlations for five variables. Notice that the correlation matrix is a symmetric matrix, that the entry in row one and column three is the same as that in row

Table 12.1 Correlation Matrix for Five Variables

Variable	1	2	3	4	5
1	1.00	.29	.68	.05	.17
2	.29	1.00	.44	.22	.03
3	.68	.44	1.00	.39	.12
4	.05	.22	.39	1.00	.41
5	.17	.03	.12	.41	1.00

three and column one. Notice also that the correlation of each variable with itself is 1.00, and this is the value entered in the principal diagonal.

EXERCISES

12.1 Given the following collection of paired measures:

X: 5 10 2 3 1 2 4 8 6 9
Y: 9 8 6 7 3 3 6 4 6 8

(a) Construct a scattergram from the data.
(b) Compute the Pearson correlation coefficient.

12.2 Given the following collection of paired measures:

X: 1 4 4 5 5 6 7 8
Y: 9 7 8 6 1 6 2 1

(a) Construct a scattergram from the data.
(b) Compute the Pearson correlation coefficient.

12.3 A social studies teacher administered an attitude scale to a group of high school students and recorded the scores of the boys separately from those of the girls. The scale purports to measure student attitudes toward "going steady," and a high score is indicative of a favorable attitude toward this activity. The scores are as follows:

Boys: 4 9 1 2 5 8 9 5 3 4
Girls: 11 8 6 7 10 8 5 11 6 8

(a) Record the sex of each student as a binary variable, awarding a one to a boy and a zero to a girl. You should then have twenty pairs of measures, each pair having a sex "score" and an attitude score. Calculate the Pearson correlation coefficient from the paired variables.
(b) What significance should be attached to the sign of the correlation coefficient?
(c) What proportion of the attitude score variance is accounted for by the sex of the respondents?

OTHER CORRELATION
COEFFICIENTS

13.1 THE SPEARMAN RANK CORRELATION COEFFICIENT

Occasionally, it will be desirable to compute a correlation coefficient when one or both variables is on an ordinal scale. There is also the situation in which the score distribution is extremely heterogeneous, and the numerical values of the scores are less meaningful than the ranks of the scores. This is the same situation as that encountered earlier when the mean was less meaningful than the median. The Spearman rank correlation coefficient is an adaptation of the Pearson product moment correlation coefficient for use with ordinal data. It may be calculated by substituting the ranks of the individuals for their scores in the Pearson formula. However, by placing the restrictions on the data that each variable will have N ranks, and that these will range from 1 to N, the formula may be simplified to:

$$r_s = 1 - \frac{6 \Sigma d_i^2}{N^3 - N}$$

The author has chosen to represent the Spearman coefficient by r_s rather than by ρ, the Greek letter rho, which is sometimes used, since he believes this is consistent with the good statistical practice of reserving Greek letters for parameters. The symbol d_i represents the difference in the two ranks assigned the ith individual.

The calculation of the Spearman coefficient from scores which have been ranked will be illustrated with the data of Fig. 12.4, where .883 has already been determined to be the value of the Pearson coefficient. To rank the data, the scores are first placed in an ordered series; it does not matter whether the highest score or the lowest score is placed first. The scores are then numbered (ranked) from 1 to N. Where tie scores are encountered, the ranks are summed and divided equally among the tied scores so that scores having the same numerical value also have the same rank. The X-variable is ranked in this fashion:

X-score: 1   2   2   3   4   5   6   8   9   10

X-rank: 1   2.5 2.5 4   5   6   7   8   9   10

The Y-variable is ranked separately, but in the same fashion:

Y-score: 2   2   3   5   5   5   6   7   7   8

Y-rank: 1.5 1.5 3   5   5   5   7   8.5 8.5 10

After the scores have been ranked, the ranks are paired in precisely the same pairings as the original scores, and the difference between the ranks is determined for each pair. This procedure is illustrated in Table 13.1.

Table 13.1   Calculation of the Spearman Rank Coefficient

X-score	Y-score	X-rank	Y-rank	d_i	d_i^2
1	2	1.0	1.5	.5	.25
2	3	2.5	3.0	.5	.25
2	2	2.5	1.5	−1.0	1.00
3	5	4.0	5.0	1.0	1.00
4	5	5.0	5.0	.0	.00
5	7	6.0	8.5	2.5	6.25
6	5	7.0	5.0	−2.0	4.00
8	6	8.0	7.0	−1.0	1.00
9	7	9.0	8.5	−.5	.25
10	8	10.0	10.0	.0	.00

$$\Sigma\, d_i^2 = 14.00$$

Once the differences have been squared and summed, the calculation of the Spearman coefficient is quite simple:

$$r = 1 - \frac{6(14)}{1000 - 10} = 1 - \frac{84}{990} = 1 - .085 = 0.915$$

This value is the Pearson correlation coefficient calculated from the ranks. It is slightly different from the earlier value of the Pearson coefficient calculated from the scores. Ordinarily, with interval data, the calculation of the correlation coefficient from scores would be preferred. However, for the individual who does not have immediate access to a calculating machine, the Spearman formula provides a good approximation of the Pearson formula with considerably less computational effort. The approximation improves with larger samples, and the Spearman coefficient is probably completely appropriate for use by the classroom teacher in the study of relationships between test scores. In the research setting, however, the Pearson coefficient will be chosen by the investigator for the study of relationships between test scores.

The Spearman rank correlation coefficient is a nonparametric statistic with certain underlying assumptions. One of these is that the measures are truly continuously distributed, and that no two persons have precisely the same quantity of a continuously distributed trait. If the precision of measurement is so gross that large numbers of tie scores are awarded, then use of the Spearman coefficient is highly questionable. If the proportion of ties is not large, its effect on the value of the coefficient is negligible.

13.2 THE POINT BISERIAL CORRELATION COEFFICIENT

The point biserial correlation coefficient is simply a Pearson correlation coefficient with one continuous variable and one dichotomous variable. The dichotomous variable is assumed to be discrete, with a uniform distribution in each of the two categories. To classify students as successful or unsuccessful (or as graduates and nongraduates) and use this as a dichotomous variable implies that all of the unsuccessful students are equally unsuccessful, and that all of the successful students are equally successful. The point biserial is often used in test item analysis, in which a correlation is calculated between the scores on a single test item and scores on the total test. The idea is to determine whether or not the individual test item is consistent with the total test. This implies that the students who answer the item incorrectly are equally ignorant of the quality measured by the item, and that the students who answer the item correctly are equally well-informed. Most statisticians agree that the point biserial is mathematically defensible, but some authorities insist that a true dichotomy must be established before the procedure is valid. The author is inclined to feel that the problem is one of precision of measurement, and the fact that the measurement is crude (that is, limited to two levels) does not invalidate the measure or its use in the calculation of a correlation coefficient.

When the point biserial is calculated with the established Pearson formulas, the dichotomized variable may be assigned any two numerical values—one and zero, or thirteen and twenty-seven, for example. The value of the coefficient is unaffected by the exact values assigned to the dichotomy; therefore the common practice is to assign values of one and zero, since this greatly reduces the computational effort. The use of a binary variable of this sort permits the derivation of a computational form which is somewhat simpler to use than the Pearson formulas, but which is mathematically equivalent:

$$r_{pbi} = \frac{M_1 - M_0}{\sigma_x} \sqrt{pq}$$

where X = the continuous variable

σ_x = the standard deviation of the continuous variable

M_1 = the X-mean of the group scoring one on the dichotomy

M_0 = the X-mean of the group scoring zero on the dichotomy

p = the proportion of the total group scoring one on the dichotomy

q = the proportion of the total group scoring zero on the dichotomy

The student's attention is directed to the fact that this formula is mathematically equivalent to the Pearson formulas given earlier only if σ_x is calculated using N rather than $N - 1$ in the denominator.

The equivalence of the Pearson and the point biserial correlation coefficients is demonstrated with the data of Table 13.2:

$$r = \frac{SP}{\sqrt{SS_x SS_y}} = \frac{-6}{\sqrt{(72)(2)}} = -0.50$$

$$r_{pbi} = \frac{M_1 - M_0}{\sigma_x} \sqrt{pq} = \frac{6.5 - 3.5}{3} \sqrt{\left(\frac{1}{2}\right)\left(\frac{1}{2}\right)} = -0.50$$

Table 13.2   Correlation of Continuous and Dichotomous Variables

X	X^2	Y	Y^2	XY
1	1	1	1	1
1	1	1	1	1
2	4	0	0	0
6	36	1	1	6
6	36	1	1	6
7	49	0	0	0
8	64	0	0	0
9	81	0	0	0
40	272	4	4	14

The student is encouraged to follow through with the calculations from beginning to end to demonstrate that the author's calculations are correct and to gain experience with the formulas.

It is completely appropriate to use nominal variables in the dichotomy, assigning a one for male and a zero for female, for example. In this case, if the females score higher on the continuous variable than the males, a negative coefficient will be obtained. Of course, whether the coefficient is negative or positive is completely arbitrary with nominal variables.

It should be noted that the magnitude of the point biserial is a function of the sizes of p and q, and it can only be as large as -1 or $+1$ when p and q are one-half. When the subjects are not evenly divided in the dichotomy, unity correlation cannot be achieved.

13.3   THE PHI COEFFICIENT

The phi coefficient is an extension of the point biserial (and the Pearson) coefficient to the situation in which both variables are dichotomized. Consider the data of Table 13.3 and the calculation of the Pearson coeffi-

Table 13.3   Correlation of Two Dichotomized Variables

X	X^2	Y	Y^2	XY
1	1	1	1	1
1	1	1	1	1
1	1	1	1	1
1	1	1	1	1
1	1	0	0	0
0	0	1	1	0
0	0	1	1	0
0	0	0	0	0
0	0	0	0	0
0	0	0	0	0
5	5	6	6	4

cient from the following data:

$$r = \frac{SP}{\sqrt{SS_x SS_y}} = \frac{1.0}{\sqrt{(2.5)(2.4)}} = \frac{1.0}{2.45} = 0.41$$

Data of this sort are quite often organized into a two-by-two bivariate frequency table, with the letters a, b, c, and d representing the frequencies of the four cells, as shown in Table 13.4.

Table 13.4   Bivariate Frequency Table for Phi Coefficient

	$X = 0$	$X = 1$
$Y = 1$	$a = 2$	$b = 4$
$Y = 0$	$c = 3$	$d = 1$

Using this notational scheme, we may calculate the phi coefficient from the cell frequencies of the bivariate frequency table by the formula:

$$\phi = \frac{bc - ad}{\sqrt{(a + b)(c + d)(a + c)(b + d)}}$$

which, in the example under consideration, is:

$$\phi = \frac{12 - 2}{\sqrt{(6)(4)(5)(5)}} = \frac{10}{24.5} = 0.41$$

The same assumptions that apply to the single dichotomy of the point biserial also apply to both dichotomies of the phi coefficient. Unity correlation can only be achieved when the subjects are evenly divided between the two categories of both dichotomies, that is, when $a + b = c + d$ and $a + c = b + d$.

13.4   THE BISERIAL CORRELATION COEFFICIENT

The biserial correlation coefficient is an index of relationship between two continuously distributed variables, one of which is treated as a dichotomy. The most common example of a continuously distributed variable which is treated as a dichotomy is the academic pass-fail variable. While academic ability is certainly a continuous variable (and probably normally distributed), pass-fail (or graduation-nongraduation) is a common dichotomous interpretation of this trait. The biserial correlation coefficient is calculated on the assumption that the dichotomized variable is in fact normally distributed. It has no direct relationship to any of the previous correlation coefficients, although it usually ranges from minus one to plus one.

The biserial correlation coefficient may be calculated from the formula:

$$r_{bi} = \frac{M_p - M_q}{\sigma_x} \left(\frac{pq}{y} \right)$$

where X = the variable recorded as continuous

σ_x = the standard deviation of the X-variable

M_p = the X-mean of the successful group

M_q = the X-mean of the unsuccessful group

p = the proportion of the subjects in the successful group

q = the proportion of the subjects in the unsuccessful group

y = the ordinate of the normal curve which divides it into two parts with p, the proportion of the total area in one part, and q, the proportion of the area in the other part

Calculated in this fashion, the biserial coefficient will be positive if the successful group has the higher X-mean and negative if the unsuccessful group has the higher mean. Table 4 in the Appendix gives the height of the normal curve for various values of p and also gives the value of pq/y.

Suppose that 60 percent of the students enrolled in a certain graduate school complete the M.A. degree and 40 percent do not complete the degree. Suppose also that the mean IQ of the students who graduate is 120, the mean of those who do not is 110, and the standard deviation is 15. The biserial correlation between IQ score and whether or not a student graduates is:

$$r_{bi} = \frac{120 - 110}{15} (.621) = .41$$

The biserial correlation coefficient has been used in almost every situation where the point biserial is used. It may be used with discriminant analysis to determine the probability of success for a student with a given score on the continuously recorded variable. This is a special application of prediction.

The biserial coefficient is not very popular with mathematically oriented statisticians, since it is doubtful that the assumption of normality of the dichotomized variable is ever established. When the assumption of normality is violated, the biserial correlation coefficient may be greater than unity.

13.5   THE TETRACHORIC CORRELATION COEFFICIENT

The tetrachoric correlation coefficient is an extension of the biserial to the situation in which two normally distributed variables are dichotomized. Its computation is quite involved; however, a good approximation of the coefficient may be achieved through the use of Table 5 in the Appendix.

The data must first be recorded in a two-by-two bivariate frequency table with a, b, c, and d being the cell frequencies. The ratio ad/bc or bc/ad (whichever is larger) is formed, and the value of the correlation coefficient is read directly from Table 5.

For example, consider the data in Table 13.4, which was previously used for the calculation of the phi coefficient. The ratio ad/bc is $\frac{2}{12}$, and the ratio bc/ad is 6. Selecting the larger of the two and turning to Table 5, we find that the tetrachoric correlation coefficient is .61.

13.6 THE CONTINGENCY COEFFICIENT

All of the correlation coefficients considered have required measurement on at least an ordinal scale or have permitted the use of nominal measurement limited to a dichotomous variable. The contingency coefficient is intended for use with nominal data recorded in bivariate frequency tables. There is no restriction upon the number of categories, provided the number of measures is quite large. The use of the contingency coefficient requires a knowledge of the chi-square statistic and will be considered in a later chapter.

EXERCISES

13.1 Two art judges ranked five oil paintings in the following fashion:

Judge I	Judge II
1 African Sunset	1 African Sunset
2 Blue Mist	2 Lily of the Valley
3 Lily of the Valley	3 The Old Fisherman
4 The Old Fisherman	4 Blue Mist
5 Country Scene	5 Country Scene

Calculate the Spearman rank correlation coefficient from the data. Are the judges in substantial agreement?

13.2 Refer to the data of Exercise 12.1. Rank the data and calculate the Spearman rank correlation coefficient. How does your answer compare with the Pearson coefficient calculated earlier with this data?

13.3 Refer to the data of Exercise 12.3. Calculate the point biserial correlation coefficient for the data. How does your answer compare with the Pearson coefficient calculated earlier with this data?

13.4 Repeat Exercise 13.3 using the biserial correlation coefficient. How does this answer compare with the point biserial coefficient? Is this an appropriate use of the biserial coefficient?

13.5 The political preference of each of a group of subjects was obtained along with an indication of whether the subject intended to vote for or against a proposed constitutional amendment. Of the group, 22 Republicans and 10 Democrats indicated they would vote for the amendment, while 8 Republicans and 16 Democrats indicated they would vote against it. Arrange the data in a bivariate frequency table and calculate the phi coefficient.

13.6 Repeat Exercise 13.5 using the tetrachoric correlation coefficient. How does your answer compare with the phi coefficient? Is this an appropriate use of the tetrachoric coefficient?

14

PRINCIPLES OF PREDICTION

14.1 THE IDEA OF REGRESSION

The terms *correlation, regression,* and *prediction* are so closely related in statistics that they are often used interchangeably. The term *correlation* refers to the degree of correspondence or relationship between two variables. Whenever two variables are correlated, it is possible, though often with a limited degree of success, to predict the score on one variable from a knowledge of the score on the other. *Regression* refers to a phenomenon characteristic of prediction with less than perfect correlation.

In 1885 Francis Galton published a paper entitled *Regression towards Mediocrity in Hereditary Stature.* It was already well established that many human characteristics are inherited—that tall people tend to have tall offspring, and that short people tend to have short offspring, etc. But Galton was interested in predicting the physical characteristics of the progeny from a knowledge of the physical characteristics of the parents. He observed that very tall parents do indeed tend to have tall offspring—but not as tall as the parents. Similarly, very short people tend to have short offspring—but not as short as the parents. This phenomenon is popularly known as *regression toward the mean.*

Consider the situation in which a student's grade point average in college is known to be correlated with rank in his high school graduating class, and the investigator is interested in predicting college grades from high school rank. Let X represent the predictor variable (high school rank) and Y the predicted variable (college grades). Given an individual whose score on the X-variable is not known, the best predicted score for this individual is the Y-mean. That is, in the long run of many such predictions to be made for many different individuals for whom there is no knowledge of the X-variable, predicting the Y-mean for each of them will give less prediction error than any other Y-score. Even when the X-variable is known, the investigator will not be able to perfectly predict the Y-variable, since the relationship between the two is imperfect. The investigator is now predicting out of partial ignorance rather than the total ignorance he experienced when he predicted the Y-score without any knowledge of the X-score. As a result, his prediction will be influenced by both his knowledge and his ignorance, and the predicted Y-score will be closer to the mean than the corresponding X-score. If the two variables are expressed as standard scores with zero mean, the absolute value of the predicted score (z_y) will be smaller than the absolute value of the predictor score (z_x). While regression toward the mean often appears to be a natural phenomenon, the modern statistician prefers to regard it as a mathematical assumption built into the statistics of prediction.

14.2   EQUATIONS AND GRAPHS OF LINEAR RELATIONSHIPS

In order for the behavioral scientist to predict scores on one variable from a knowledge of scores on another, he must also have a knowledge of the relationship between the two. This means he must first study a group of subjects on whom both measures are available before he can extend his findings to subjects on whom only one measure is available. The bulk of this chapter will be devoted to the group on whom both measures are available, although we are primarily interested in being able to predict the second variable from a knowledge of the first.

One very important assumption underlying the prediction process considered here is that the relationship between the two variables is linear and can be represented by a straight line drawn on a scattergram. The relationship between two linearly correlated variables may also be represented by the equation of a straight line, which is in the general form of $Y = bX + c$.

Figure 14.1 is a graph of the particular linear equation, $Y = 2X - 3$. The b in the general form of the equation has been replaced by the constant 2 in this particular equation, and c has been replaced by -3. There is a value of Y for every possible value of X in this equation, which is another way of saying that there is an infinity of solutions to the equation. All of these solutions are represented by the line in Fig. 14.1.

The constant b is known as the *slope of the line*, which is the rate at which Y changes with change in X. Its value may be determined from the graph of the line by selecting any two values of X (call them X_1 and X_2)

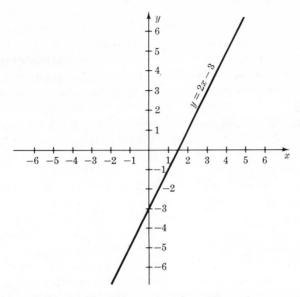

Figure 14.1   Graph of a Linear Equation.

plus the corresponding values of Y (call them Y_1 and Y_2), then calculating the slope from:

$$b = \frac{Y_2 - Y_1}{X_2 - X_1}$$

The slope of the line also happens to be the tangent of the angle which the line makes with the horizontal axis. The constant c is called the Y-intercept, and it may be determined from the graph as that point at which the line crosses the vertical axis. It is the value of Y that corresponds to $X = 0$.

14.3   THE LEAST SQUARES CRITERION

If two variables are perfectly correlated, a scattergram may be constructed and a straight line drawn through *all* of the points on the graph. When predicting from this situation, one need only identify a point on the line which is directly above the X-score on the horizontal axis and identify the corresponding Y-score on the vertical axis. More often, the investigator will determine the equation of the line and solve for Y by substituting the value of X in the equation.

Of course, it is highly unlikely that the behavioral scientist will ever encounter perfect correlation in a practical prediction situation. He can anticipate that the points on the scattergram will not all fall on a straight line, and he must seek to identify a straight line which best expresses the relationship between the two variables. He is interested in arriving at a linear prediction equation which gives the least amount of error when predicting the second measure from the first i.e., for that group of individuals on whom both measures are known. The assumption is made that this equation will also give the least amount of error when predicting the second measure for individuals on whom only the first measure is known.

Let X represent the first score and Y the second score of the group on whom we have both scores. We will seek a linear equation with which we may predict Y from X with the least amount of error. The predicted scores will lie on the straight line, but many of the Y-scores previously recorded from actual measurements will not lie on this line. We must make a distinction between the Y-scores previously recorded and the Y-scores predicted from the equation. Let $\hat{Y}$ represent the predicted scores (those which all lie on the prediction line). To the extent that these differ from the recorded Y-scores, they contain an error component. The difference between the predicted score ($\hat{Y}$) and the recorded score (Y) for a given predictor score (X) is error and will be represented by the lower case letter e:

$$e = Y - \hat{Y}$$

Thus, if the recorded Y-score for a given individual is smaller than the predicted score, the error is negative; if the recorded Y-score is larger than the predicted score, the error is positive. The situation is illustrated in Fig. 14.2. The data were borrowed from Fig. 12.3, for which we have already calculated the sums of squares, sum of products, and the correlation coefficient. The author has also determined the appropriate prediction equation and has drawn the prediction line. The predicted score for an individual whose X-score is 5 is also 5 in this particular case. Notice, however, that the one individual who scored 5 on the X-variable has a score of 7 on the Y-variable. The prediction error for this individual is 2. Of course, in the typical research setting, where the number of subjects is much larger, several subjects would probably have an X-score of 5, but there is no reason to expect that each would have the same recorded Y-score. Each, however, would receive the same predicted Y-score ($\hat{Y}$), and the error would vary from person to person with some errors being positive and some negative.

Now, how does one determine the prediction equation that gives the least error? There is an infinite number of prediction equations for a given set of data in which the algebraic sum of the errors is zero. In fact, if the

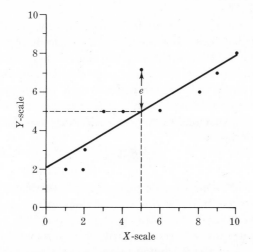

Figure 14.2   Linear Prediction Error.

point on the scattergram that corresponds to both the X- and the Y-mean is located, any straight line drawn through this point will be a prediction line for which the sum of the errors is zero. Obviously, the solution does not lie in having a prediction equation with large amounts of negative error canceled by equally large amounts of positive error. Another solution might call for the use of the absolute values of the errors, but this is a mathematical dead end; thus we seek a solution that permits additional manipulation of the findings. The answer is quite simple—it is the one we encountered in selecting the variance over the mean deviation as a measure of dispersion. We shall square the errors and choose as our prediction equation that equation for which the sum of the squared errors ($\Sigma\ e_i^2$) is smaller than that for any other equation. This is called the *least squares criterion*, and its derivation requires the use of differential calculus; however, the problem has already been solved, and we shall be able to determine the prediction equation through the use of simple algebra. This prediction equation is more often referred to as a *regression equation* and its locus as a *regression line*.

14.4   REGRESSION EQUATIONS FOR STANDARD SCORES

One of the consequences of regression toward the mean is that the regression line will pass through a single point corresponding to the means of the Y-, $\hat{Y}$-, and X-distributions. Whenever scores are expressed as deviations from the mean (as in the case of $x_i = X_i - M$ or $z_i = x_i/\sigma$), the mean of the distribution of the deviation scores is zero. If the regression equation is calculated with deviation scores, there will be a point on the regression line at which both x and y (or z_x and z_y) equal zero. In other words, the y-intercept will be zero and will drop out of the equation.

The least squares solution yields a regression equation for use with z-scores having a slope equal to the correlation between the two variables, and the equation becomes:

$$z_y = r_{xy}z_x$$

If the two sets of z-scores are perfectly linearly correlated, r_{xy} will equal unity, and the regression line will lie at a 45-degree angle to the horizontal axis. If there is no relationship between the two variables, r_{xy} will be zero, and the regression line will coincide with the horizontal axis, which is also the y-mean, and the predicted score for all subjects is the y-mean. If the two variables are correlated, but imperfectly, the regression line will intercept the horizontal axis at some angle between 0 and 45 degrees.

Figure 14.3 shows the prediction line for the situation in which the correlation is .67 between two sets of z-scores. Notice the regression toward the y-mean; an individual with $z_x = +3$ is awarded $\hat{z}_y = +2$, while an individual with $z_x = -3$ is awarded $\hat{z}_y = -2$.

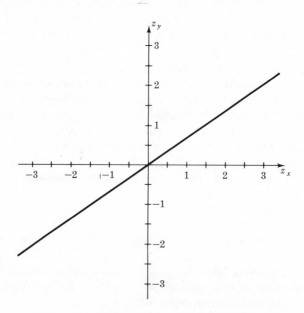

Figure 14.3   Regression with Standard Scores.

14.5 REGRESSION EQUATIONS FOR DEVIATION SCORES

The general form of the regression equation for deviation scores is:

$$\hat{y} = bx \qquad \text{where } x = X - M$$

The value of b may be derived by substituting into the regression equation for z-scores and by the application of some algebra:

$$b = \frac{SP}{SS_x} \quad = \Sigma x y \; - \text{Sums of Products}$$

The use of deviation scores will simplify the problems of regression analysis when a single variable is predicted from a knowledge of several others. This is called *multiple regression* and will be considered in one of the advanced chapters of this text.

14.6 REGRESSION EQUATIONS FOR RAW SCORES

The general form of the regression equation for raw scores is:

$$\hat{Y} = bX + c$$

The only distinction between deviation scores and raw scores is the placement on the score scale (the shifting of the mean). Thus, the value of the slope is the same as for the deviation scores:

$$b = \frac{SP}{SS_x}$$

The value of the Y-intercept is chosen so that the Y-mean corresponds to the X-mean (so that they fall at the same point on the regression line):

$$c = \bar{Y} - b\bar{X}$$

Consider the data of Fig. 14.2, for which the following results were previously determined: $\bar{X} = 5$, $\bar{Y} = 5$, $SS_x = 90$, $SS_y = 40$, and $SP = 53$. The regression equation may be determined as follows:

$$b = \tfrac{53}{90} = .59$$
$$c = 5 - .59(5) = 2.05$$
$$\hat{Y} = .59X + 2.05$$

The regression line of Fig. 14.2 was constructed with a knowledge of the regression equation. This equation may now be used to predict the $\hat{Y}$-score for any individual for whom the X-score is known. For example, for an individual whose X-score is 2:

$$\hat{Y} = .59(2) + 2.05 = 3.23$$

The student should refer to Fig. 14.2 and verify that this is in fact the $\hat{Y}$-score that corresponds to $X = 2$ on the regression line. He should also note that neither of the two persons in the original distribution who scored 2 on the X-variable has this Y-score.

The problem of errors of prediction will be considered next.

When one variable is predicted from another, it is customary to refer to the predictor variable as the *independent* variable and the predicted variable as the *dependent* variable.

14.7   THE STANDARD ERROR OF ESTIMATE

Let us return for a moment to the situation in which the investigator has no knowledge of the independent variable. Here the predicted score will be the Y-mean, and 68 percent of the subjects will fall within one standard deviation of the Y-mean if the population is normally distributed. Thus, the investigator may predict the Y-mean for all subjects and have the assurance that 68 percent of his predictions will be accurate within plus or minus one standard deviation. This is a perfectly worthless piece of information, but it does illustrate the important idea of the standard deviation as an estimate of error.

The error of prediction for a given subject was previously defined as the difference between his actual score on the dependent variable and his predicted score on the dependent variable; that is, $e = Y - \hat{Y}$.

Of course, knowledge of the magnitude of this error after the final score is in is not a very useful piece of information in the practical prediction situation. What is needed is an estimate of this error at the time of prediction—when the score on the dependent variable has not yet been awarded to this particular subject. The investigator will require an accurate estimate of the amount of error involved in any given prediction situation. One such estimate is the *standard error of estimate*, which may be defined as the standard deviation of the errors of prediction:

$$SE_{est} = \sqrt{\frac{\Sigma (Y - \hat{Y})^2}{N - 2}} = \sqrt{\frac{\Sigma e^2}{N - 2}}$$

As a general rule, it will be considerably more convenient to calculate the standard error of estimate from the following mathematically equivalent formula:

$$SE_{est} = S_y \sqrt{1 - r^2} \quad \text{where } S_y = \sqrt{\frac{SS_y}{N - 2}}$$

The standard error of estimate, of course, must be calculated from that group on whom both the independent and dependent measures are available, but it is intended for use with those individuals on whom the dependent variable is not yet available. The student has now entered into the study of statistical inference. The individuals from whom the regression equation was calculated are the sample; the population consists of those individuals for whom the dependent variable might be predicted using the regression equation. The investigator is interested in extending his findings from the sample to the population and in having an expression of the confidence with which he can do this. Notice that the formula for the standard deviation of the errors is calculated with $N - 2$ in the denominator. This term, $N - 2$, is the degrees of freedom of the statistic, a rather involved topic which will be considered in greater detail in the chapters which follow. Its use, as with the use of $N - 1$ before, is to insure an unbiased estimate of the population parameter.

The calculation of the standard error of estimate is illustrated with the data of Fig. 14.2:

$$SE_{est} = \sqrt{\frac{40}{8}} \sqrt{1 - .78} = 1.05$$

This suggests that 68 percent of the individuals for whom a given predicted score is calculated will in fact score within plus or minus 1.05 score points of the predicted score. Referring to Fig. 14.2, we note that seven of the ten subjects from whom the regression equation was calculated did in fact score within 1.05 points of their predicted scores. There are certain assumptions underlying the use of the standard error of estimate in this fashion. First, the group from whom the regression equation was calculated must be genuinely representative of the group with whom the regression equation is used for purposes of prediction. Second, the errors of prediction are assumed to be normally distributed. Third, the errors of prediction are equally distributed at all points on the regression line. This last condition is known as *homoscedasticity*, and its violation usually increases the errors of prediction for extreme scores. This is not usually a problem in practical prediction, since the possible success or failure of the extreme individuals is more easily predicted than that of the individuals in the middle of the distribution. In other words, there may be considerable error in the predicted scores for the extreme cases, but this is likely to be of little practical consequence to the behavioral scientist and should not prohibit the use of the standard error of estimate.

EXERCISES

14.1   Given the following collection of scores, in which the X-variable is IQ score and the Y-variable is the grade received in a certain course of study:

i	X_i	Y_i	X_i^2	Y_i^2	X_iY_i
1	120	4	14400	16	480
2	110	3	12100	9	330
3	100	2	10000	4	200
4	90	0	8100	0	0
5	80	1	6400	1	80
	500	10	51000	30	1090

(a)   Make a scattergram of the scores recorded above.
(b)   Calculate the correlation between the two variables.
(c)   Determine the slope of the regression line.
(d)   Determine the Y-intercept of the regression line.
(e)   Write the regression equation.
(f)   Draw the regression line on the scattergram.
(g)   Calculate the predicted score for each of the X-scores recorded above.
(h)   Calculate $\Sigma \, e^2$.
(i)   Notice that three of the points on the scattergram lie on a straight line. The equation of this line is $Y = .1X - 8$. Draw the line on the scattergram and calculate $\Sigma \, e^2$ for this prediction line.
(j)   Compare the answers to (h) and (i) above.
(k)   Determine the standard error or estimate.
(l)   What is the predicted grade for an individual with an IQ of 95?

15

STATISTICAL
PROCEDURES IN
MEASUREMENT

15.1 INTRODUCTION

Measurement and research in the behavioral sciences are intimately related—good research is invariably dependent upon measurement. By the same token, certain research is essential to the development of a good measurement device. Most of the measurement devices used by behavioral scientists can be classified as tests, and there are three essential attributes of a good test: *standardization, validity,* and *reliability.*

Standardization means that the test has been administered to a well-defined group and that careful records have been kept with respect to their performance. These records are called *norms,* and the well-defined group is the *norm group.* The selection of an appropriate norm group is important, because it serves as the basis for the determination of test validity and reliability and because it will often be the point of reference for interpretation of measures derived from other subjects. To facilitate the process of comparison, standardization will generally include the development of one or more kinds of standard scores.

A test is valid when it measures what it is intended to measure. The important idea here is the use which is to be made of the test scores. A test may be valid for one use, but this is no assurance that it is valid for other uses. For example, an IQ test might be a valid measure of scholastic aptitude, but it might be completely invalid as a measure of executive ability.

Reliability refers to the precision of measurement. It is an essential ingredient in validity; a test cannot be very valid if it is not very reliable. However, a test may be reliable without being valid. The professional test constructor will seek to express reliability and validity in mathematical terms. Summarizing, validity is an expression of the extent to which a test measures what it is supposed to measure, reliability of how well it measures whatever it measures.

The personal judgment of the particular individual who scores a test should not be a factor affecting the score. To the extent that several qualified persons can score a test for a given individual and award the same score, a test is said to be objective; to the extent that several different scorers do not award the same score, there is an error component in the score. *Objectivity* is an essential ingredient in reliability. Because a lack of objectivity is reflected in a lack of reliability, it will not usually be necessary to seek a mathematical definition of objectivity.

15.2   THEORY OF RELIABILITY

Man has never been able to demonstrate that any measurement on a continuous scale is without error, and all of the available evidence suggests that every such measurement has some error in it. The logical conclusion is that every measurement on a continuous scale consists of two components, the true value plus some measurement error. This idea may be expressed mathematically by:

$$x_i = t_i + e_i$$

where x_i = an attempt to measure something (the observed value)
t_i = the true value
e_i = the error component in the measurement

If we can assume that an error of measurement is as likely to be positive as it is to be negative, it is logical to conclude that the algebraic sum and the mean of the errors would both be zero if an infinite number of attempts to measure the same thing were made. The definition of the true score, then, is the mean of an infinite number of attempts to measure the same thing.

This idea may be extended to the situation in which one measure is taken on each of several subjects, so that the true score is a variable. Then the following relationship may be defined:

$$\sigma_x^2 = \sigma_t^2 + \sigma_e^2$$

Reliability may be defined as the ratio of the true score variance to the observed score variance. The symbol r_{xx} will be used to represent the coefficient of reliability, a quantity which exists only in theory and which is sometimes called the *self-correlation:*

$$r_{xx} = \frac{\sigma_t^2}{\sigma_x^2} = 1 - \frac{\sigma_e^2}{\sigma_x^2}$$

There are several different ways to estimate the coefficient of reliability, and they are distinguished from each other by defining error in slightly different ways.

It can be demonstrated that the coefficient of reliability is a function of the number of items in a test—the greater the number of items in a test, the more reliable the test tends to be. Thus, a shortened form of an established test can be expected to be less reliable than the original. The so-called Spearman-Brown prophecy formula is an expression of the relationship between test length and reliability:

$$r_e = \frac{k r_{xx}}{1 + (k - 1) r_{xx}}$$

where r_{xx} = is the reliability of the original test

r_e = is the expected reliability of the test with the number of items changed

k = is the ratio of the number of items in the changed test to the number in the original

When the length of the test is doubled, the formula reduces to:

$$r_e = \frac{2 r_{xx}}{1 + r_{xx}}$$

15.3 TEST-RETEST RELIABILITY

The most obvious method for determining reliability of a test calls for administering it to the same sample on two different occasions, then defining reliability as the Pearson product moment correlation between the two sets of scores. This method assumes that it is practical and valid to administer the test to the same group of persons twice in a relatively short period of time. To some persons, this is the only true reliability—to others, it is only one way of getting at reliability, and perhaps not the best way for a given instrument.

Test-retest reliability is sometimes called a *coefficient of stability* and carries with it the idea that the trait being measured is fairly stable through time—at least through the period of time that separates the two administrations. It also implies that the second score is not greatly affected by the double exposure.

It is not a very practical procedure for measurements of traits that are believed to be in a state of flux or traits that are being subjected to undue influence during the time between administrations. These factors would tend to decrease the estimate of reliability. And it is not a very practical approach to reliability where remembering an answer or calculation or procedure might affect the second score. These factors would tend to give a spurious increase in the estimated reliability.

Error in test-retest reliability is anything that leads a person to get different scores on the two administrations.

15.4 EQUIVALENT FORMS RELIABILITY

One way of solving some of the problems associated with test-retest reliability is to have two equivalent forms of the same instrument. Both forms are administered to the same group of subjects, and the estimated reliability is simply the Pearson product moment correlation between the two sets of scores. The two administrations may take place quite close in time, and there is little danger of error introduced by change through time or due to practice effects.

However, to the extent that the two forms are not equivalent, a new source of error is introduced. As a general rule, this approach to reliability will yield a somewhat lower estimate of reliability than test-retest. There is also the problem of generating two equivalent forms of the same instrument, a task that may not be economically or otherwise feasible.

Error in this procedure is also anything that leads a person to get different scores on the two administrations, but this is not exactly the same error as in the test-retest procedure and may not lead to the same coefficient of reliability.

15.5 SPLIT-HALF RELIABILITY

One of the approaches to reliability most popular with the test constructor is the split-half technique. A single administration of the instrument is made, the test split into two halves which are scored separately, and a Pearson correlation coefficient between the two scores is calculated. Then the Spearman-Brown prophecy formula is used to compensate for the fact that the reliability was estimated from a test one-half the length of the final form. For example, suppose the correlation between the two halves were .80, then the expected reliability of the full test would be:

$$r_e = \frac{2(.80)}{1 + .80} = .89$$

While a number of ways of carrying out a split-half reliability have been devised, the best is probably the odd-even approach, in which the odd-numbered test items are treated as one-half of the test and the even-numbered as the alternate half. This ensures that approximately the same amount of time will be devoted to both halves and helps to insure some equivalence of the two forms if there is a tendency for the items to become more difficult near the end of the test.

This type of reliability is one form of *internal consistency* reliability. It is not very meaningful in speeded tests, and to the extent that the test is speeded, the coefficient of reliability is inflated. Whenever a test has a time limit, and some of the subjects do not complete all items, the score is as much a function of the number of items completed as it is of the number of items answered correctly. There is a tendency in this situation for the items in the first part of the test to be answered correctly and those in the last part to be answered not at all (which is interpreted as answering incorrectly). The correlation between the odd and even items of such a test will be very high.

There are many desirable characteristics in split-half reliability: the conditions of administration are near identical for both halves, there is little or no time lag between administrations; and the attitudes or response sets of the respondents should be near identical for the two administrations. But some researchers suggest that this is not reliability at all, and it certainly does tend to yield a higher estimate of the reliability than test-retest. There is no dispute that split-half reliability is completely inappropriate where a substantial speed component is measured.

15.6 RELIABILITY BASED ON ITEM STATISTICS

Kuder and Richardson developed their approaches to reliability based on item statistics in an attempt to overcome some of the deficiencies of split-half reliability. Their formulas split a test into as many parts as there are items in the test. Assumptions are made that all of the items measure a single trait and that each respondent attempts every item. The basic formula is Kuder-Richardson formula 20:

$$r_{xx} = \left(\frac{n}{n - 1}\right)\left(\frac{\sigma^2 - \Sigma\, pq}{\sigma^2}\right)$$

where n = the number of items in the test

σ^2 = the variance of the test scores

p = the proportion of the subjects answering a given question right

$q = 1 - p$

Some computational effort can be saved through the use of Kuder-Richardson formula 21, if one can assume that all of the items are of equal difficulty:

$$r_{xx} = \frac{n\sigma^2 - M(n - M)}{(n - 1)\sigma^2}$$

Notice that this reliability estimate may be calculated with only a knowledge of the mean and variance of the scores plus the number of items on the test.

To the extent that a test measures on more than one dimension (and this is often desirable to get maximum validity), the Kuder-Richardson formulas will underestimate the reliability. To the extent that the test is speeded, they will overestimate the reliability.

15.7 THE STANDARD ERROR OF MEASUREMENT

Beginning with the mathematical definition of reliability given earlier, one may derive a formula for the standard deviation of the errors of measurement. This is more often referred to as the *standard error of measurement:*

$$SE_m = \sigma_e = \sigma_x \sqrt{1 - r_{xx}}$$

This is usually interpreted as an indication of the probable extent of error in the score of a given individual. Suppose an IQ test has a reliability of .96 and standard deviation of 15:

$$SE_m = 15 \sqrt{1 - .96} = 3$$

Thus, if one can assume that the errors of measurement are normally distributed about a given score and equally distributed throughout the score range, 68 percent of the subjects will be regarded as having a true IQ score within 3 points of their IQ scores as determined from the test.

15.8 THE PROBLEM OF VALIDITY

In the physical sciences both direct and indirect measurement are encountered. The weight of an object is measured directly by placing it on a balance and comparing it to a standard. Temperature, however, is measured indirectly. When the physical scientist says he is measuring temperature, it is more probable that he is measuring the length of a column of mercury and inferring from it the temperature. He does not measure heat directly—he measures not what temperature is, but what it does.

The same idea holds true with respect to most measurement by the behavioral scientist. First, he must carefully assign a definition to the trait under study—not in terms of what it is, but what it does. He asks the question, how does one behave if one possesses this trait? Or, what behavior distinguishes an individual with much of this trait from one with little of it? For example, he might define intelligence as the ability to solve certain kinds of problems. This is called an *operational definition*; the trait, itself, is not observable, but it is defined in terms of observable phenomena. The behavioral scientist then measures the observable phenomena and draws conclusions with respect to the unobservable trait. Of course, there is always the danger that he is not measuring what he proposes to measure at all. This is the problem of validity—to what extent is he measuring what he intends to measure?

15.9 CONTENT VALIDITY

Content validity is demonstrated by showing how well the content of a test samples the situations about which conclusions are to be drawn. It is especially important for achievement and proficiency measures.

This is sometimes referred to as logical or sampling validity, or validity by definition. It usually implies the use of some expertise to define a universe of interest, the careful drawing of a representative sample of ideas from this universe, and the preparation of test items that match these ideas. Some empirical method will be used to demonstrate the reliability of the instrument, but empirical methods may not necessarily be required to establish validity.

Statistical methods may be applied to the problem of content validity. For example, the test constructor may perform an analysis of the test items which includes a point biserial (or biserial) correlation between the score on each item and the score on the total test. The assumption is that the necessary expertise is available to identify intelligently what the total test is all about. Test items that are not consistent with the total are either revised or eliminated.

15.10 CRITERION-RELATED VALIDITY

Criterion-related validity is demonstrated by correlating test scores with one or more external criteria. Since few direct measures are available to the behavioral scientist, the big problem is to find a criterion. Consider, however, a test intended to measure scholastic aptitude; in this case, the criterion would be the degree of success in a scholastic setting. Such an approach to validity is essentially a matter of prediction—something which behavioral scientists are able to do with only a limited degree of success. Nonetheless, this is usually regarded as the best approach to validity when a defensible criterion can be identified. Criterion-related validity is sometimes called *empirical* validity. The coefficient of validity is the correlation between the test and the external criterion; one should not expect it to be as high as the coefficient of reliability.

Concurrent validity is a cross-sectional approach to the problem of prediction. Instead of studying the future success of the subject, scores of already successful or unsuccessful people are used as a standard of comparison. Thus, one might validate a test of mechanical aptitude by securing scores of mechanics and nonmechanics to demonstrate that the test can discriminate between the two. This is sometimes called *discriminant* validity; it is not as highly regarded as predictive validity. In the final analysis, the most useful test is one that can be used to predict human behavior, and concurrent validity is not direct evidence of this quality.

15.11 CONSTRUCT VALIDITY

Construct validity is not a simple idea. It includes both logical and empirical approaches to validity, as do all other approaches, but it seeks a unique combination of the two. The approach is as follows, in a general sense, but the exact techniques will vary greatly:

(1) The investigator sets forth a proposition that a test measures on a certain dimension, or designs a test to measure that dimension.
(2) He then inserts that proposition into current theory about that dimension.
(3) He hypothesizes empirical relationships based on the theory.
(4) Then a study of the relationships is carried out using the test to measure the dimension.

If the findings are consistent, some support is lent to the measurement and the theoretical framework within which it is presented. For example, the author constructed an instrument that purports to measure certain personal values. Another investigator hypothesized that psychologists who subscribe to different psychological theories are people who differ on certain personal values. The investigator's conclusions were borne out by research, lending support to both the validity of the instrument and the theory of the investigator.

INTRODUCTION TO
RESEARCH

16.1 SCIENCE AND THE SCIENTIST

We live today in an age where science is a dominant force in our lives. We are members of a scientific society. Our individual and collective stature as members of a profession is irrevocably associated with the understanding and use of scientific knowledge and methods.

There is no universally accepted, inflexible, unchanging definition of science. It is a many-sided human enterprise that has been defined in many different ways. The author will attempt to define science in a fashion which is consistent with the purposes and content of this textbook. His definition is intended to simplify the problems of organizing, carrying on, and interpreting research in the behavioral sciences.

The central purpose of science is to provide an objective, factual account of the universe in which we live. Science seeks an empirical and verifiable explanation of natural phenomena in contradistinction to artistic, philosophic, and religious ways of explaining things.

There are two fundamental assumptions which underlie all scientific endeavor: (1) the behavior of the universe is orderly, it is not capricious, chaotic, or spontaneous; (2) every natural event has an explanation which may be discovered by intelligent and diligent people. There is a third assumption which most—but not all—scientists make: truth is independent of the observer; i.e., when two scientists disagree about some natural phenomena, the usual conclusion is that at least one of the two is wrong. Still, most scientists like to hold all, or almost all, of their conclusions as tentative and subject to later revision.

Science is an activity carried on in a specific way to advance human knowledge. It is not the knowledge itself, since neither having possession of scientific knowledge nor applying it determines one to be a scientist. It is the contribution to that knowledge that is science and determines one to be a scientist. Science, then, is the activity of empirical research—systematic study, using observation and experimentation to establish new facts or principles. Reduced to its simplest form, a form consistent with the purposes and content of this text, science is the systematic study of the relationships between variables. Usually, this activity takes place in a restricted setting with the intent to draw generalizations beyond that setting.

Far too much emphasis has been placed on something called the scientific method; there are many different ways of getting at the truth, not just one. More important than a scientific method is a scientific attitude. The scientist does not seek to settle disputes by oratory, or by violence, or even by taking a vote. The scientific attitude is "Let's try it and see." In other words, the scientist seeks the solution to life's problems through research. To the extent that the individual learns to live by the spirit of science, he shares in the liberation of the human intellect.

16.2 RESEARCH VARIABLES

We have just equated science and research and have defined these as the systematic study of relationships between variables. The variables that enter into research are classified in two broad categories: independent and dependent. In the simplest research (and often the most profound), only two variables are studied, one independent and one dependent. The investigator is interested in determining whether or not a relationship exists between the two variables, and if a relationship does indeed exist, what the nature of that relationship is. Suppose an educator wishes to determine whether the traditional textbook or a new programmed textbook is more effective in teaching algebra. He might arrange to teach one group of students from the traditional textbook and another from the new programmed textbook. At the conclusion of the learning experience, an examination would be given to determine how well the students learned their algebra. Then a determination would be made as to whether one group learned more than the other, and if so, which group, and perhaps, how much more. In this piece of research, the textbooks used were the independent variable, and the examination was the dependent variable.

The terms *independent variable* and *dependent variable* are extremely difficult to define in a way that fits all situations. However, in a given research setting they are usually quite readily identified. We will define the two in terms of a *presumed* cause-and-effect relationship. Variations in the independent variable are presumed to cause variations in the dependent variable. Often, it is not practical to establish a cause-and-effect relationship, and it may not be the intent of the investigator to establish one, but this is a convenient way of classifying the variables. Another way of looking at the distinction between the two is that determination of an individual's measure on the independent variable will ordinarily *precede* the determination of his measure on the dependent variable.

In the prediction situation, the score on the independent variable is used to predict the score on the dependent variable. In an experimental setting, the independent variable is manipulated by the experimenter, and the dependent variable is the *criterion* by which he determines whether or not anything of consequence happen when (not because) he manipulated the independent variable. However, much important research is conducted in which the investigator only observes and explains the independent variable and does not manipulate it. Suppose the educator previously mentioned had been interested in the relationship between the sex of the student and achievement in the study of algebra. In this case, the sex of the student is the independent variable.

Let us now suppose that this same educator is primarily interested in comparing the relative effectiveness of the two different textbooks, but he suspects that other variables, such as IQ and sex of the student, may also influence the dependent variable and confound the outcome of the experiment. This suggests that the independent variables should be broken into two groups: (1) the experimental variables which the investigator is interested in manipulating to determine their influence upon the dependent variable, and (2) other variables he wishes to control to prevent their influence upon the dependent variable. The problems of organizing, carrying out, and interpreting behavioral research center around these three kinds of variables: (1) independent variables whose effects are to be studied, (2) independent variables whose effects are to be controlled, and (3) dependent variables which are observed in order to determine the consequences.

16.3 STATEMENT OF THE RESEARCH PROBLEM

Practically all empirical research begins with questions to be answered, problems to be solved, or ideas to be tested. It is highly improbable that the investigator will succeed in his quest for a solution unless the problem is clearly stated in a form that lends itself to research. One simply cannot solve a problem unless he knows what the problem is. The problem must be stated in such a fashion that it can be solved, that the solution will be recognized when it is found, and that the solution will be defensible.

The problem may be stated as a hypothesis—a conjectural statement about the relationships that might exist among two or more variables. The investigator may choose to make a forthright statement of conviction, suggesting that a particular relationship is believed to exist. Or he may choose to state the problem in the form of a *null* hypothesis—a statement that no significant relationship between the variables exists. The usual intent in this case is to disprove the hypothesis; it is stated in this form since it is a convenient approach to statistical analysis. The beginning researcher may find that the statement of a null hypothesis assists him to clearly state the problem in a fashion that lends itself to statistical treatment. The author, however, is partial to the statement of the problem in the form of a question to be answered. Typically, the research question asks what relationship, if any, exists between these variables.

Whatever form is chosen for the statement of the problem, it is not ready for research until it is expressed as the study of relationships between variables. The variables must be clearly specified in a fashion that leaves no doubt as to the manner in which they will be assessed. Similarly, the nature of the relationships to be studied must be clearly identified. Thus far, the student has been exposed only to simple correlational methods of studying these relationships, but he will find that the remainder of this text is devoted to techniques for handling increasingly more complex relationships.

The identification and quantification of the variables in behavioral research is crucial. Such terms as creativity, intelligence, self-actualization, authoritarianism, anxiety, and success are completely worthless in the research setting unless they are operationally defined. An operational definition is a definition that assigns meaning to an idea by specifying observable phenomena which represent the idea. Operational definitions are simply instructions for assessing human behavior—for translating constructs into researchable variables. Occasionally, an operational definition is an easy thing to come by. For example, it might be completely appropriate in a given piece of research to define intelligence as a score on a certain IQ test, or to define academic success as graduation from college. However, the development of good operational definitions can be one of the most difficult problems encountered by the behavioral researcher. The task is complicated by the fact that many of the terms used by practitioners in the behavioral sciences are so vague that they defy definition in operational terms. The very survival of some of these terms is due to the fact that they cannot be operationally defined and empirically verified. However, there are many important ideas in the behavioral sciences that are still in process of development and that may eventually lend themselves to research verification. These should not be cast aside just because a simple operational definition is not immediately forthcoming. The eventual quantification of these is one of the most important tasks confronting the behavioral researcher.

The problem of relationships is essentially one of experimental design, which is discussed in the following section.

16.4 EXPERIMENTAL DESIGN

It is customary to think of an experiment as a contrived situation divorced from the complexities of life—a situation in which the researcher manipulates a single variable and observes its effect on a second variable while all other variables are held constant. Although this type of research has been regarded as the most highly developed and formalized of all scientific endeavor, it has certain disadvantages in the study of human behavior. First, some of the most important variables of interest to the behavioral scientist do not lend themselves to experimental manipulation. For example, the socio-economic background of an individual may be a critical determinant of his behavior in a race riot. These variables are of great interest to the behavioral scientist, but he is hardly in a position to manipulate them in a laboratory setting. Second, human behavior is exceedingly complex, involving a great deal of interaction with a multitude of variables, some of which are unobserved. This environment cannot be adequately duplicated in the laboratory; consequently, behavioral research conducted in the laboratory is often a sterile endeavor. Whether a child has been reared in riches or in poverty, or has lived in a democracy or under a dictatorship for the first ten years of his life is likely to be a more important determinant of his behavior in a given situation than any treatment administered by the researcher. Most classroom research, for example, demonstrates that the achievement of the student is determined much more by the experiences he had before he attended a given class than the experiences he entered into while in the class. The author is not opposed to laboratory research in the behavioral sciences; he is merely suggesting that the term *experiment* has a much broader meaning than that which is usually accorded it. Some of the most valuable research in the behavioral sciences takes place without the intervention of the investigator, who merely observes the greatest "experiment" of all, life itself. His role as a scientist is a creative one, nonetheless. He must develop hypotheses based on previous observations, he must identify and quantify the relevant variables, he must establish the relationships between these variables, and he must interpret these relationships in a fashion that makes a significant contribution to our knowledge about human behavior.

The fundamental principles of experimental design are given in the following list:

(1)   The major purpose of an experiment is to describe the effect of the independent variable on the dependent variable.

(2)   The term "independent variable" refers to any induced or selected variation in the experimental methods or materials whose effect is to be observed and evaluated.

(3)   Evaluation of the effects of the independent variable necessitates a *criterion*—the dependent variable. The outcome of the experiment will be described in terms of variation in the dependent variable. This will most often be a prediction equation, a comparison of means or medians of the various experimental groups, or a comparison of the frequency of occurrence of certain events.

(4)   Provision is made for controlling the effects of other variables whose presence might influence the variation of the dependent variable.

The simplest type of experimental design is the simple randomized design with independent samples. Each experimental treatment is administered to a different set of subjects drawn at random from the parent population. At the conclusion of the experiment all subjects are evaluated using the dependent variable, and the investigator then determines whether the different experimental treatments have produced different results. The effects of irrelevant variables are controlled by the random assignment of subjects, since there is equal probability that a subject of given ability will be assigned to a particular treatment group.

Some improvement in the simple randomized design can be accomplished through the use of representative random samples. Suppose that a given population is known to be distributed with 25 percent of the individuals having an IQ above 110, 50 percent in the range 90 to 110, and 25 percent below 90, and suppose that IQ is believed to be related to the dependent variable. Stratified random samples could be drawn, each of which had an IQ distribution similar to that of the parent population, and the investigator could be reasonably sure that he is dealing with groups of approximately equal ability. This design has the additional advantage of yielding separate results for each of the various levels of ability, if this is desired.

Occasionally, the behavioral scientist will be interested in studying simultaneously the effects of two (or more) independent variables on a given dependent variable. This is equivalent to carrying on two experiments at the same time with one collection of experimental subjects. Suppose that a psychiatrist is interested in studying concomitantly the effects of a certain drug and electroshock therapy. He might randomly assign one-fourth of the available patients to receive both the drug and electroshock, one-fourth to receive only the drug, one-fourth to receive only electroshock, and one-fourth to receive neither. It is immediately apparent that one-half of the total number of subjects receives the drug, and one-half do not. Similarly, one-half of the subjects receive electroshock, and one-half do not. In addition to the economies effected by carrying on two experiments at the same time with one group of subjects, the investigator has the additional opportunity of being able to study possible *interaction* between the two independent variables. Interaction is a treatment effect that results from two variables working together. This type of design, along with the one using stratified samples, will be treated in detail in the chapter on two-dimensional analysis of variance.

Sometimes it is possible to conduct an experiment in which all treatments may be successively administered to each subject. This assumes that the effects of each treatment are unaffected by the fact that one or more treatments have been previously administered to the same subjects, a condition that is not likely to be met where learning is taking place. The effects of irrelevant variables are controlled by the fact that the various experimental groups have precisely the same subjects. A closely related design calls for matching the subjects on one or more variables before assigning them to the various experimental treatments. Suppose the investigator is interested in determining which of two instructional methods is more effective with mentally retarded children. The children are first grouped in pairs, each pair containing two children as much alike in learning ability as possible. Then one member of each pair is randomly assigned to each of the two experimental treatments. Experimental designs of this sort are called *matched-pairs* or *related samples designs*.

One of the most useful designs in educational research involves the *analysis of covariance*. It has already been noted that the characteristics that a student brings into an experiment are likely to have greater impact on the dependent variable than any experimental variables to which he might be subjected. However, to the extent that these characteristics can be identified and quantified, they can be used to predict the dependent variable, and their contribution to its variance can be statistically eliminated from the final analysis. The control of several variables in this fashion requires the use of *multiple regression*.

In summary, the design of an experiment requires that the investigator enumerate the variables to be studied, classify them as independent or dependent, indicate which variables are to be controlled and in what manner, and describe the relationship to be examined.

16.5 STATISTICAL ANALYSIS

Closely related to the design of the experiment is the matter of choosing an appropriate statistical procedure to complete the analysis. The choice of statistical procedure will be based upon the following considerations:

(1)   The level of measurement. Practically all statistical analyses are classified according to the level of measurement achieved in the dependent variable, some procedures being appropriate for use with nominal variables, others with ordinal variables, and others with interval and ratio variables.

(2)   The distribution of the measures. Some procedures require continuous distribution of the dependent variable, others are suited for discrete data. Some procedures require normality of distribution, others make no such restriction.

(3)   The number of samples. Generally, these can be classified as procedures appropriate for use with one sample, two samples, and two or more samples.

(4)   Whether the samples are related or unrelated. Some statistical procedures are suited for use with independent samples, others for use in the situation where the measures on the dependent variable for the various samples are related to each other. These measures may be related to each other either because they are taken from the same subjects or from subjects matched before being assigned to the samples.

16.6 RESEARCH AND PROBABILITY

There is a common misconception that science deals with certainties. Nothing could be further from the truth. Science does not deal with certainties, it deals with probabilities. The best the scientist can hope to draw from his research is that, given a specified level of an independent variable and well-defined control over certain other variables, it is extremely probable that certain effects will be observed in the dependent variable. The mathematics of probability enable him to quantify the level of confidence he has in his prediction.

17

BASIC PROBABILITY
THEORY

17.1 DEFINITION OF PROBABILITY

By way of introduction to probability theory, let us first review the terms *population* and *sample*. A *population* is a collection of objects, events, or individuals having some common characteristic which the researcher is interested in studying. A *sample* is a smaller group chosen from the population to participate directly in the research as a representative of the population. Invariably, the use of probability theory to interpret the research will prescribe the use of a *random* sample. This is a sample chosen in a fashion which insures that every object, event, or individual in the population has equal chance of being drawn for the sample. The important idea is that the selection of each object in the sample is a chance event.

It is very difficult to arrive at a definition of the term *probability* that is both completely acceptable to the mathematician and readily communicable to the nonmathematician. However, the following definition is relatively simple and quite adequate for the use of probability theory in the behavioral sciences: The probability that an event (object or person) will occur is equal to the relative frequency of occurrence of that event (object or person) in the population under consideration.

For example, suppose the population consists of 100 persons who have previously been determined to include 30 females and 70 males. The probability of selecting a male in random drawings of a single individual from this group is 70/100. Similarly, the probability of drawing a female is 30/100. Turning to a different kind of sampling, let us suppose that when a student guesses the answer to an objective question on an examination his answer is a completely random one. Then the probability of guessing the right answer to a true-false question is 1/2, and the probability of guessing the right answer to a multiple-choice question with five responses is 1/5. In these last examples, the probability of the event is its relative frequency in a hypothetical frequency distribution, or *long run* frequency distribution as it is sometimes called. When statisticians say the probability of heads in the flip of a coin is one-half, they mean the relative frequency of heads is 0.50 in a distribution consisting of a very large number of flips of the coin.

17.2 SEVEN FUNDAMENTAL RULES OF PROBABILITY

Rule 1 The probability of occurrence of any one of a set of equally likely events is one divided by the number of events. This may be stated symbolically by $P(e_i) = 1/N$.

For example, given a group of six boys, whose names are Joe, John, Robert, Jim, Elmer, and George. The probability of selecting Elmer in a single random draw is 1/6.

Rule 2 If a frequency distribution exists, and if there is more than one object or event in a given class, the probability of drawing an object of this particular class is equal to the class frequency divided by the number of objects; that is, $P(e_i) = f_i/N$.

For example, given a frequency distribution consisting of 20 Englishmen, 30 Canadians, and 5 Australians, the probability of selecting an Australian in a single random draw from this population is 5/55, which is 1/11.

Rule 3 The probability of any event cannot be less than zero (there are no negative probabilities) or greater than one (one represents a sure thing). Stated mathematically, $0 \leqslant P(e_i) \leqslant 1$.

Rule 4 The sum of the probabilities of all of the possible events in a population must equal one. This may be stated symbolically by $\Sigma P(e_i) = 1$.

This rule has important implications in sampling procedures. It indicates, for example, that one cannot draw samples from a trichotomous population (such as the Englishmen, Canadians, and Australians above), cast aside those individuals in one of the categories, and treat the remaining individuals as though they were drawn from a dichotomous population.

Rule 5 The probability of an event occurring plus the probability of that event not occurring is equal to one. This is usually stated symbolically by $q = 1 - p$, where p represents the probability of occurrence and q represents the probability of nonoccurrence.

This rule suggests that any population may be dichotomized and treated with statistical procedures appropriate for use with dichotomous populations. (The population considered above could be regarded as consisting of Canadians and non-Canadians, for example.)

Rule 6   The addition rule of probability The probability that any one of a set of mutually exclusive events will occur is the sum of the probabilities of the separate events.

When two events are mutually exclusive, if one occurs, the other cannot occur. For example, in an objective type of examination, the student's answer to a question is either right or wrong; it cannot be both.

Refer to the population containing 20 Englishmen, 30 Canadians, and 5 Australians. The probability of an Englishman is 20/55, and the probability of a Canadian is 30/55. The probability that a single random selection will produce either an Englishman or a Canadian is 50/55, the sum of the probabilities of the separate events.

Rule 7   The multiplication rule of probability The probability that a combination of independent events will occur is the product of the separate probabilities of the events.

When two events are independent, the occurrence or nonoccurrence of one has no effect on the other.

For example, the probability that a student will guess the correct answer to the first question on a true-false examination is 1/2, and the probability that he will guess the correct answer to the second question is also 1/2. But before he answers any questions, the probability that he will guess the correct answers to the first two questions is 1/4, the product of the separate probabilities. Similarly, the probability that he will guess the correct answers to the first three questions is 1/8, and to the first four questions 1/16. Again, we are assuming that his answers are completely random.

17.3 SAMPLING FROM A FINITE POPULATION WITHOUT REPLACEMENT

A special problem arises when sampling from populations sufficiently small that the drawing of a sample appreciably changes the distribution of the population. Suppose a population consists of six men and three women. The probability of drawing a male from this population is 6/9 or 2/3. However, if a male is drawn and not replaced, the population becomes five men and three women, and the probability of drawing a second male is 5/8. Thus, the probability of drawing two males in a row without replacing the first is 5/12, the product of the separate probabilities. Similarly, the probability of drawing three males in a row without replacement from a population consisting of six males and three females is: 6/9 × 5/8 × 4/7, which is 5/21. The behavioral scientist is unlikely to draw samples of this sort, but he should be aware of the implications if he encounters such a procedure.

17.4 SAMPLING FROM A FINITE POPULATION WITH REPLACEMENT

The problem of sampling from a small population is appreciably simplified if the sampling is done with replacement. As before, the probability of drawing a male from a population containing six males and three females is 2/3. If a male is drawn and replaced, then the probability of drawing a male the second time is also 2/3, and the probability of drawing two males in a row with replacement is 4/9. The probability of drawing three males in a row with replacement is 8/27.

17.5 SAMPLING FROM VERY LARGE POPULATIONS

Most of the sampling procedures considered in this text assume that the researcher is drawing samples from a population so large that sampling does not appreciably change the distribution of the population. This is equivalent to sampling with replacement. As a general rule, behavioral scientists will draw their samples from very large populations, and these populations will be regarded as approaching infinite size.

17.6 FACTORIAL NOTATION

Consider the situation in which a high school principal is asked to rank five teachers in order of their teaching abilities. He may choose the first teacher from any one of the five, but has only four from which to choose the second, three from which to choose the third, two from which to choose the fourth, and one for the fifth. This is a problem in sampling without replacement, and if the principal's selection is a purely random one, the probability of any particular order of the five teachers is 1/5 × 1/4 × 1/3 × 1/2 × 1/1, which is 1/120.

Calculating the product of the number of choices available to the principal yields 5 × 4 × 3 × 2 × 1, which is 120, the number of possible ways in which the five teachers could be ranked. If the selection were completely random, each of these arrangements would be equally probable, each having a probability of 1/120.

Calculations like these occur often enough in probability problems to warrant a special type of notation, called *factorial* notation: $n!$ (read "*n*-factorial") is the product of all of the integers from n to 1. For example, 3! is 3 × 2 × 1, which is 6, which is the number of ways three different objects can be arranged in order. Suppose the three objects were represented by the letters A, B, and C. They can be arranged in order six ways, as follows: ABC, ACB, BAC, BCA, CAB, and CBA. In the same fashion, one may determine the number of ways six things can be arranged in order: $6! = 6 \times 5 \times 4 \times 3 \times 2 \times 1 = 720$.

It is customary to let both 1! and 0! equal one. The logic of this will become clear to the student as he studies the following sections and works some examples.

17.7 PERMUTATIONS

Suppose a principal had been asked to choose and rank the best three teachers out of a group of five. He would have had five from which to choose the first, four from which to choose the second, and three from which to choose the third, yielding a total of 60 ways in which three teachers can be chosen and ranked from a group of five. A group of objects chosen and ordered in this fashion is called a *permutation*. The number of permutations of n objects taken r at a time is:

$$P(n, r) = \frac{n!}{(n - r)!}$$

In the example quoted above, the value of n was 5 and the value of r was 3:

$$P(5, 3) = \frac{5!}{(5 - 3)!} = \frac{5!}{2!} = \frac{5 \times 4 \times 3 \times 2 \times 1}{2 \times 1} = 60$$

In a completely random selection, each of the possible permutations is equally probable.

17.8 COMBINATIONS

Suppose the principal were requested simply to choose the three best teachers from a group of five without ranking the three. The possible ways of ranking three objects is 3!, which is 6. This suggests that the number of combinations of five things taken three at a time (without respect to order) is equal to the number of permutations divided by 3!, that is, 60/6, which is 10. The general equation for determining the number of combinations possible from n objects taken r at a time is:

$$C(n, r) = \frac{n!}{r!(n-r)!}$$

In a completely random situation, each of these possible combinations is equally probable.

The symbol $\binom{n}{r}$ is also used to represent the number of combinations of n objects taken r at a time.

EXERCISES

17.1 A group of students is known to contain 15 students of superior ability, 40 students of average ability, and 12 students of below average ability.
   (a) What is the probability of a random choice yielding a student of superior ability? $\frac{15}{67}$
   (b) What is the probability of a random choice yielding either a student of superior ability or a student of below average ability? $\frac{27}{67}$

17.2 A true-false examination contains eight questions.
   (a) What is the probability of a student guessing the right answer to any given question? Assume a completely random selection. $\frac{1}{2}$
   (b) What is the probability of a student getting all of the questions right simply by guessing? $\frac{1}{2}^8 = \frac{1}{256}$
   (c) What is the probability of a student getting all of the questions wrong simply by guessing? $\frac{1}{2}^8 = \frac{1}{256}$

17.3 A multiple-choice examination contains five questions, each with three responses.
   (a) What is the probability of a student guessing the right answer to any given question? $\frac{1}{3}$

(b)   What is the probability of a student guessing a wrong answer to any given question? $2/3$

(c)   What is the probability of a student getting all of the questions right simply by guessing? $\frac{1}{3}^5 = \frac{1}{243}$

17.4   A college student claims he can identify four different brands of cigarettes by taste. An experiment is set up to test his ability, he is blindfolded and given four cigarettes (one of each brand), one at a time, and asked to identify them by brand name.

(a)   How many different ways can four different cigarettes be presented to him one at a time? $4! = 24$

(b)   What is the probability that he will correctly identify all four brands simply by guessing? $\frac{1}{24}$

17.5   A graduate student is given the choice of answering any four questions from a group of seven on an essay examination. How many different options does he have; that is, how many groups of four questions can be selected from a collection of seven? $\frac{7!}{4!}$

17.6   Eight coeds will compete in a beauty contest, and the judges will award trophies to the first-, second-, and third-place winners.

$\frac{8!}{(8-3)!} = 336$

(a)   How many ways can three winners be ranked out of a field of eight competitors?

(b)   Suppose some college boys place the names of the eight girls in a hat and attempt to duplicate the judges' selection by drawing names from the hat. What is the probability that the first draw from the hat will be the first-place winner? What is the probability that three names drawn from the hat would be the first-, second-, and third-place winners, in that order?

$\frac{1}{8}$

$\frac{1}{8} \times \frac{1}{7} \times \frac{1}{6} = \frac{1}{336}$

18

THE BINOMIAL
PROBABILITY
DISTRIBUTION

18.1 APPLICATION OF THE FUNDAMENTAL RULES
OF PROBABILITY

Consider the situation in which a student who has not studied his lessons is confronted with a "pop-quiz" containing five multiple-choice questions each having three possible responses. Let us assume that his answers are random guesses, so that the probability of his guessing the correct answer to any given question is $1/3$ and the probability of his guessing a wrong answer to any given question is $2/3$.

The probability that the student will guess all five correct answers is determined by applying the multiplication rule of probability: $1/3 \times 1/3 \times 1/3 \times 1/3 \times 1/3 = (1/3)^5$, which is $1/243$.

The probability that the student will get four right answers and one wrong is a bit more complicated. First, the number of combinations of five things taken four at a time is five. This should be verified from the combinatorial formula given in Section 17.8. Each of these five possible combinations of four right answers and one wrong is equally probable. Now, let us consider any one of these equally probable combinations. The probability that he will get a particular set of four questions right is $1/3 \times 1/3 \times 1/3 \times 1/3$, and the probability that he will get the other question wrong is $2/3$. Applying the multiplication and addition rules of probability, we see that the probability that the student will get four answers right and one wrong is $5(1/3)^4(2/3)$, which is $10/243$. (The five equally probable combinations are added together by multiplying by five.)

Now consider the situation in which the student gets three answers right and two wrong. There are ten equally probable combinations of three

124

right and two wrong answers to five questions. Applying the multiplication and addition rules as before, we find that the probability that the student will get three right and two wrong is $10(1/3)^3(2/3)^2$, which is 40/243.

Similarly, the probability that the student will get two right and three wrong may be determined to be $10(1/3)^2(2/3)^3$, which is 80/243. And the probability he will get one right and four wrong is $5(1/3)^1(2/3)^4$, which also happens to be 80/243. Finally, the probability of getting all five questions wrong is $1(1/3)^0(2/3)^5$, which is 32/243. Now, consider the total picture:

$$P(5 \text{ right, } 0 \text{ wrong}) = 1(1/3)^5(2/3)^0 = 1/243$$
$$P(4 \text{ right, } 1 \text{ wrong}) = 5(1/3)^4(2/3)^1 = 10/243$$
$$P(3 \text{ right, } 2 \text{ wrong}) = 10(1/3)^3(2/3)^2 = 40/243$$
$$P(2 \text{ right, } 3 \text{ wrong}) = 10(1/3)^2(2/3)^3 = 80/243$$
$$P(1 \text{ right, } 4 \text{ wrong}) = 5(1/3)^1(2/3)^4 = 80/243$$
$$P(0 \text{ right, } 5 \text{ wrong}) = 1(1/3)^0(2/3)^5 = 32/243$$
$$\text{sum of the probabilities} = \overline{243/243} = 1$$

Let us take note of some of the relationships evident in these calculations. The first multiplier, which we will later refer to as the coefficient, is the number of combinations of right answers possible for a given score. It may be calculated using the combinatorial formula, where n is the number of questions and r is the number of right answers. Notice that the coefficients are symmetrical, that the first and last coefficients in the distribution are equal to one, that the second and next to last are equal to n. The quantity in the first parentheses is the probability (p) of getting a right answer, and the exponent is the number of right answers (r). Similarly, the quantity in the second parentheses (q, or $1 - p$) is the probability of getting a wrong answer, and the exponent is the number of wrong answers ($n - r$). Calculations of this sort are greatly facilitated through the use of the binomial probability distribution equation.

Finally, take note of the fact that the sum of all of the probabilities in the distribution is equal to one. This is an example of the use of the addition rule of probability, which may be extended to any combination of probabilities in the distribution. For instance, the probability of a student getting four or more questions right is equal to the probability of getting four right plus the probability of getting five right, which is 11/243.

18.2 BINOMIAL POPULATIONS

Many dichotomous variables are encountered in the behavioral sciences: for example, male versus female sex, right versus wrong answers to test questions. Many other nominal variables are usefully dichotomized by classifying all objects in a given population as either belonging or not belonging to a given classification: for example, psychologists and non-psychologists, delinquents and nondelinquents. The reader is reminded that these categories must be mutually exclusive and collectively exhaustive. One cannot dichotomize a population by classifying people as Republicans and Democrats, discarding any individuals who do not fit in one of these two classes. Occasionally, it will be desirable to reduce a continuous variable to a dichotomy: for example, classifying students as above or below average ability.

Populations that are split into two mutually exclusive and collectively exhaustive categories are called *binomial* populations. The binomial probability distribution is a useful device for studying binomial populations.

18.3 THE GENERAL EQUATION OF THE BINOMIAL PROBABILITY DISTRIBUTION

The general equation of the binomial probability distribution (BPD) is given by:

$$P(r \text{ in } n) = \binom{n}{r} p^r q^{n-r}$$

where n = the size of a random sample of events

r = the number of events in the first category

$n - r$ = the number of events in the second category

p = the probability of any single event falling in the first category

$q = 1 - p$, which is the probability of any single event falling in the second category.

For a given problem, only r is a variable, and r may vary from zero to n.

It will not be necessary to calculate the values of the binomial coefficients—the first multiplier in the equation—since these have been conveniently tabled for small values of n in Table 6 in the Appendix. For larger values of n, the binomial equation becomes unwieldy, and another procedure (to be presented later in this chapter) will be substituted.

18.4 APPLICATION OF THE BINOMIAL PROBABILITY DISTRIBUTION

To make use of the BPD, every object in the population under study must be clearly identifiable as belonging to one of two classes, although the distribution of these objects (the number in each class) may not be known. Second, the researcher must have some basis for arriving at the values of p and q, the probabilities associated with membership in the two classes. The BPD is a *theoretical* probability distribution—a mathematical model for certain kinds of random events. In addition to its use to determine the probability of certain kinds of random events (such as chance scores on objective tests), the BPD may also be used as a model for testing certain kinds of research hypotheses. In the latter application, a sample distribution is compared to the theoretical distribution to see if the sample deviates significantly from the theory. Both applications will be illustrated.

Suppose a teacher constructs a ten-item true-false examination and is interested in determining what kinds of scores students might be expected to get by guessing at the answers. Referring to the general equation of the BPD, we need only substitute the appropriate values to determine the probability of a student getting any particular score, assuming his responses are random. For this example, $n = 10$, p and q are both $1/2$, and r will range from 0 to 10. The probability associated with each score from 0 to 10 is presented in Table 18.1.

Table 18.1   Distribution of Chance Scores to Ten T-F Questions

r	$\binom{n}{r}$	p^r	q^{n-r}	Probability
10	1	$(1/2)^{10}$	$(1/2)^0$	$1/1024 = .001$
9	10	$(1/2)^9$	$(1/2)^1$	$10/1024 = .010$
8	45	$(1/2)^8$	$(1/2)^2$	$45/1024 = .044$
7	120	$(1/2)^7$	$(1/2)^3$	$120/1024 = .117$
6	210	$(1/2)^6$	$(1/2)^4$	$210/1024 = .205$
5	252	$(1/2)^5$	$(1/2)^5$	$252/1024 = .246$
4	210	$(1/2)^4$	$(1/2)^6$	$210/1024 = .205$
3	120	$(1/2)^3$	$(1/2)^7$	$120/1024 = .117$
2	45	$(1/2)^2$	$(1/2)^8$	$45/1024 = .044$
1	10	$(1/2)^1$	$(1/2)^9$	$10/1024 = .010$
0	1	$(1/2)^0$	$(1/2)^{10}$	$1/1024 = .001$

sum = 1.000

Let us now suppose that the teacher has decided that a passing grade on this examination is achieved by every student who correctly answers seven or more questions. What is the probability of a student guessing seven or more correct answers? The addition rule of probability applies, and one need only add the probabilities associated with scores of seven or more correct answers to determine that the answer is .172. In other words, the student has about one chance in six of getting seven or more correct answers if his answers are completely random.

Now, let us revise the test, using multiple-choice questions with four responses to each question. The theoretical distribution is given in Table 18.2.

Table 18.2   Distribution of Chance Scores to 10 M-C Questions

r	$\dbinom{n}{r}$	p^r	q^{n-r}	Probability
10	1	$(1/4)^{10}$	$(3/4)^0$	$1/1048576 = .000$
9	10	$(1/4)^9$	$(3/4)^1$	$30/1048576 = .000$
8	45	$(1/4)^8$	$(3/4)^2$	$405/1048576 = .000$
7	120	$(1/4)^7$	$(3/4)^3$	$3240/1048576 = .003$
6	210	$(1/4)^6$	$(3/4)^4$	$17010/1048576 = .016$
5	252	$(1/4)^5$	$(3/4)^5$	$61236/1048576 = .058$
4	210	$(1/4)^4$	$(3/4)^6$	$153090/1048576 = .146$
3	120	$(1/4)^3$	$(3/4)^7$	$262440/1048576 = .250$
2	45	$(1/4)^2$	$(3/4)^8$	$295245/1048576 = .281$
1	10	$(1/4)^1$	$(3/4)^9$	$196830/1048576 = .188$
0	1	$(1/4)^0$	$(3/4)^{10}$	$59049/1048576 = .056$

sum $= .998$

Notice that the probability of seven or more correct answers is now .003. In other words, the student has about one chance in 300 of getting a passing grade if his responses are completely random. Of course, the chance factor in the examination could be reduced even more by using five-response multiple-choice questions or by using more questions.

It was indicated earlier that the binomial probability distribution may also be used for testing certain kinds of research hypotheses. We shall illustrate the idea with an experiment in extrasensory perception (ESP). Suppose that a certain individual claims to have special powers of this sort, and a psychologist agrees to conduct a brief experiment to test these powers. The supposedly gifted person is blindfolded and placed in a chair with his back to a table. Three objects are placed on the table: a book, a bouquet of flowers, and a reading lamp. The psychologist points to one of these objects and asks the subject to tell which of the three it is. If the subject gave five correct answers in six attempts, how would you assess the outcome of the experiment? Consider the binomial probability distribution of Table 18.3.

Table 18.3   Chance Distribution for ESP Experiment

r	$\binom{n}{r}$	p^r	q^{n-r}	Probability
6	1	$(1/3)^6$	$(2/3)^0$	$1/729 = .001$
5	6	$(1/3)^5$	$(2/3)^1$	$12/729 = .016$
4	15	$(1/3)^4$	$(2/3)^2$	$60/729 = .072$
3	20	$(1/3)^3$	$(2/3)^3$	$160/729 = .219$
2	15	$(1/3)^2$	$(2/3)^4$	$240/729 = .329$
1	6	$(1/3)^1$	$(2/3)^5$	$192/729 = .263$
0	1	$(1/3)^0$	$(2/3)^6$	$64/729 = .088$

$$\text{sum} = .998$$

The probability that the subject would get five or more correct responses is .017 or about one chance in sixty, unless he has something more than chance going for him. Of course, a much more satisfactory piece of research could have been conducted simply by increasing the number of trials (the sample size) from six to twenty, for example.

The student's attention is directed to the fact that the use of the binomial will not always require the calculation of the entire distribution as given in the three illustrations. There is no reason, for example, why the analysis of the ESP experiment could not have been completed by calculating only the probabilities for $r = 6$ and $r = 5$.

18.5   THE NORMAL APPROXIMATION OF THE BINOMIAL

The reader's attention is directed to Figs. 10.1 and 10.2, which include both the histogram and frequency polygon for the chance distribution of responses to two true-false examinations. Notice the close resemblance of the frequency polygons, especially that for the ten-item examination, to the normal curve. It can be demonstrated mathematically and empirically (that is, by establishing an empirical sampling distribution) that the binomial distribution approaches the normal distribution if the sample size (in this case, the number of test items) is made very large. This suggests the use of the normal distribution as an approximation of the binomial distribution. The probabilities associated with normal distributions are conveniently tabled, and the binomial becomes extremely awkward to work with as sample size increases.

The accuracy of the normal approximation of the binomial is dependent upon the sample size and the value of p. If p is equal to or close to .50, the approximation is quite good for samples as small as 10 in size. However, as p departs from .50, larger samples are required to achieve a good approximation. Research has established that an approximation acceptable for most purposes is achieved when the products np and nq are both equal to or greater than 5. Thus, if p equals .50, n should be 10 or larger; if p or q is as small as .20, then n should be at least 25.

The mean of any binomial distribution is equal to the product np. This is immediately apparent in a symmetrical distribution (where $p = .50$), but is easily demonstrated to be true for other binomial distributions as well. With a little more effort it can be demonstrated that the variance of any binomial distribution is equal to the product npq. Thus, a z-score may be calculated, substituting r for the score, np for the mean, and npq for the variance:

$$z = \frac{r - np}{\sqrt{npq}}$$

If two such z-scores are calculated and entered in a table of areas under the normal curve, the area between the two is the probability associated with all of the scores in that score range. It is imperative that the real limits of scores be used when treating areas as probabilities. For example, to determine the probability of a score of 5, one would determine the area between two z-scores, one corresponding to a score of 4.5 and the other to 5.5. To determine the probability of a score of 5 or larger, one would determine the area to the right of 4.5, and to determine the probability of a score of 5 or smaller, one would determine the area to the left of 5.5. The use of real limits in this fashion is essential whenever a continuously distributed model is used to approximate a discrete probability distribution. This is sometimes referred to as a *continuity correction*, or a *Yates correction*, from the man who originated it.

The use of the normal approximation of the binomial will be illustrated with the chance distribution of ten true-false items—a distribution for which the exact probabilities have been calculated and reported in Table 18.1. Suppose the investigator wishes to determine the probability of a score of 5. Then:

$$z = \frac{4.5 - 5}{1.58} = -.316$$

$$z = \frac{5.5 - 5}{1.58} = +.316$$

Rounding the z-scores to two decimal places, and entering these in Table 3 of the Appendix, we find that the area between the two scores is determined to be .251 (that is, .1225 plus .1225). This is a reasonable approximation of the .246 recorded in Table 18.1. As a matter of fact, the error in the approximation is mostly due to rounding of the z-scores.

Now, let us use the normal approximation to determine the probability of a score of 7 or more. Only one z-score will be needed in this case, since we are interested in all of the area to the right of the score.

$$z = \frac{6.5 - 5}{1.58} = .95$$

Again referring to Table 3 in the Appendix, we find the area to the right of the score is .171 (from .5000 minus .3289), which is an excellent approximation of the exact probability calculated earlier.

EXERCISES

18.1 Given a true-false examination with six questions:
   (a) What is the probability of a student guessing all six correct answers if his responses are completely random?
   (b) What is the probability of a student giving random responses and scoring none right?
   (c) What is the probability of a student giving random responses and scoring exactly five right?
   (d) What is the probability of a student giving random responses and scoring five or more?

18.2 Repeat Exercise 18.1 for a multiple-choice examination having three responses for each question.

18.3 Refer to the ESP experiment reported in Section 18.4. Suppose the number of trials was increased to ten; what then is the probability of the subject getting eight or more right by chance?

18.4 Refer again to the ESP experiment. Suppose the number of trials was held at six, but the number of objects on the table was increased to four; what then is the probability of the subject getting five or more right by chance?

18.5 Given a 100-item true-false examination:
   (a) What is the probability of a student scoring 70 or higher if all of his responses are completely random?
   (b) What is the probability of a student scoring 60 or higher if all of his responses are completely random?
   (c) What is the probability of a student scoring 50 or higher if all of his responses are completely random?

18.6.   Given a 50-item multiple-choice examination in which each question has
four responses:
  (a)   What is the probability of a student scoring 35 or higher if all of his
        responses are completely random?
  (b)   What is the probability of a student scoring 30 or higher if all of his
        responses are completely random?
  (c)   What is the probability of a student scoring 25 or higher if all of his
        responses are completely random?

18.7   Which examination has the least chance score factor, the 100-item true-
false examination or the 50-item multiple-choice examination?

BASIC SAMPLING THEORY

19.1 RANDOM SAMPLING

Typically, scientific research involves intensive study of small groups (called *samples*) in order to draw conclusions about much larger groups (called *populations*). As a general rule, this is a necessary economy in any kind of empirical research, and its validity rests upon the assumption that the samples are genuinely representative of the populations. It is important for the student to understand that the researcher works with samples, but his primary concern is with conclusions that apply to populations.

A population is a collection of objects, events, or individuals having some common characteristic that the researcher is interested in studying. A sample is a smaller group of objects, events, or individuals selected from the population for actual participation in the research. For example, a behavioral scientist might be interested in analyzing the study habits of freshmen students at a large university. It is unlikely that he can collect meaningful data on all of the individuals in the population. He must select a sample to represent the total population and analyze the study habits of this sample.

Statistical inference consists of an accumulation of techniques for drawing inferences or generalizations from samples to populations. Such inferences from samples to populations will always be subject to some error, since each sample from a given population might be expected to have somewhat different characteristics and yield somewhat different results in the experimental setting. Statistical inference uses the mathematics of probability to provide means for estimating this error, that is, for stating quantitatively how much confidence can be placed in the outcome of the experiment.

Invariably, statistical inference will require that the researcher use random samples. A simple random sample is a sample chosen from a population in such a fashion as to insure that every object, event, or individual in the population has equal chance of being drawn for the sample. There are several methods of accomplishing this end, all of which require that a listing be made of every object, event, or individual in the population. For a relatively small population, it is conceivable that the researcher might achieve a random sample by listing all of the objects in the population on separate slips of paper, placing these in a container, then forming the sample by drawing slips from the container. However, most research will be facilitated by using more sophisticated techniques for drawing random samples.

One of the more popular procedures for compiling random samples calls for the use of a table of numbers generated by a random process. Table 1 in the Appendix is such a table. This particular table is unusually compact for a table of random numbers, but it is quite adequate for drawing random samples as large as several hundred from populations consisting of as many as several thousand objects. Its use is outlined as follows:

(1)   All objects in the population are numbered from one to N.

(2)   The investigator selects a row number and a column number in the table as a point from which to start drawing his sample. The method of selecting row and column numbers should approximate a random process; the particular method to be used is left to the imagination of the reader.

(3)   Beginning at the point selected above, and using as many columns as are necessary to include numbers as large as N, the investigator systematically draws a list of numbers from the table. He may move to the left or right, upward or downward from his starting point.

(4)   The list of numbers drawn from the table is then referred to the population listing. Objects in the population whose numbers correspond to the numbers drawn from the table are designated for the sample.

Suppose the population consists of 500 subjects, and the investigator wishes to draw a random sample of size 30. Suppose also that Column 14, Row 9 is selected as the starting point. In order to include numbers as large as 500 in the sample, the investigator will read columns 14–16. Thus, the first number from the table is 918, but this is too large, so it is cast aside. Proceeding downward from the starting point, the investigator would draw the following sample:

354	297	096	256	293	109	044	411	425	307
175	206	478	482	416	399	213	059	496	485
055	147	446	229	476	215	113	406	325	172

The drawing of a random sample is often impractical in actual research situations. The investigator is often confronted with the fact that he cannot choose whom he wishes to participate in his research, and he must make use of the available subjects. The population of interest may be too large—all adolescent females in the world, or all fourth grade students of arithmetic, for example. The investigator must work with an available group, one which he hopes will not differ significantly from the population of interest on any relevant variable. For example, the results of research with polio vaccines in the United States are readily generalizable to Germany, where the population speaks a different language, to Africa, where the population is of a different race, and to China, where the population has a different social structure. By the same token, the educational researcher may be quite safe in extending the results of learning research with fourth graders in Colorado to fourth graders in Wisconsin. There is always danger of error in extending the results of research in this fashion, just as there is always danger involved in driving an automobile down a crowded highway. The problem is one of being both prudent and productive. A special problem exists for the classroom researcher. He seldom has any authority over the manner in which students are assigned to the various classes, but he often feels that the procedures used to assign students closely approximate random procedures. This is probably an unwarranted assumption, but it may enable him to conduct an otherwise impractical piece of research. He must recognize that he makes use of the techniques of statistical inference at considerable risk of violating this underlying assumption, and it becomes his responsibility to defend the validity of his samples with whatever logic he can muster. Sometimes, the only solution is to suggest the existence of a hypothetical population from which these subjects might be regarded as a random sample.

Quite often in research, an available group of subjects is randomly divided into samples. For example, an available group of 40 subjects might be randomly divided into an experimental group with 25 subjects and a control group with 15 subjects. The use of inferential statistics with such samples is completely valid; however, the population to which inferences may be drawn is rather restricted.

The author likes to draw a distinction between inferences and generalizations. The investigator is privileged to draw inferences only to that population from which he has drawn a random sample. He may draw generalizations to other individuals whose differences from the research sample are such that they cannot reasonably be expected to interact with the experimental variables.

19.2   SAMPLING DISTRIBUTIONS

It has already been noted that the value of a statistic can be expected to vary from one sample to another, even though the samples are all randomly selected from the same parent population. Imagine a very large population—so large that the drawing of many samples from it will not change its distribution. Suppose we draw a random sample of size 30 from this population and determine the value of some statistic (the mean, for example) from this sample. Then, let us repeat the process 1000 times, producing 1000 random samples of size 30 and the corresponding 1000 sample statistics. A frequency distribution may be established for these 1000 sample means, just as a frequency distribution may be established for a collection of scores. This is what is meant by an *empirical sampling distribution*. From a knowledge of mathematical statistics, one could derive a theoretical model for the distribution of this statistic, just as we derived a theoretical model for the frequency distribution of certain binomial populations in a previous chapter. This model is called the *theoretical sampling distribution*, or simply the *sampling distribution*, of the statistic. It is the frequency distribution of an infinite number of statistics, each calculated from a separate random sample of the same size from the same parent population. The theoretical sampling distribution of every common statistic has been derived and is available for use in testing hypotheses about random samples.

When the theoretical sampling distribution of a statistic is known, the investigator may readily determine the relative frequency with which different sample values of this statistic are expected to occur, provided that these samples are randomly selected from a parent population for which the value of the parameter is specified. Of course, these relative frequencies may be directly equated with the probability of occurrence of different sample values of this statistic. Thus, by drawing a random sample and calculating the value of the statistic, it may be compared to the theoretical sampling distribution, and the probability that it was drawn from a population with a specified value of the parameter may be determined. Similarly, one may draw two or more random samples and determine the probability that they were drawn from the same parent population. This is the essence of statistical hypothesis testing.

The *expected value* of any statistic—the predicted value which would give the least error (see Sections 14.1 and 14.3) for many samples—is the mean of the sampling distribution. The *standard error* of any statistic is the standard deviation of its sampling distribution. The expected value and standard error of theoretical frequency distributions are parameters and will be represented by Greek letters.

19.3   THE SAMPLING DISTRIBUTION OF THE MEANS

The sampling distribution of the means is of particular interest to the behavioral scientist, since this statistic is typical of most of the samples used in behavioral research and is most often used to represent these samples in statistical analysis.

It can be shown that the expected value of the sample means is equal to the population mean regardless of the form of distribution of the population and the size of sample drawn. In other words, the sample mean is an unbiased estimate of the population mean and will be used to estimate the population mean when the value of the parameter is unknown.

If the scores of the individuals in the parent population are normally distributed, the sampling distribution of the means is also normal. Given any population (with any distribution, normal or otherwise) with mean μ and finite variance σ^2, as the sample size increases without limit, the sampling distribution of the means approaches a normal distribution with mean μ and variance σ^2/N. This is known as the *central limit theorem*, and its importance lies in the fact that it enables the researcher to use the familiar normal probability distribution for testing hypotheses about populations having any form of distribution. The size of sample needed to take advantage of the central limit theorem depends upon the extent to which the population under study deviates from normality. Generally, the sampling distribution of the means closely approximates the normal for samples of 100 or larger, no matter how radically the parent population deviates from the normal. For most behavioral research, samples of size 30 will be adequate to insure close approximation of the normal by the sampling distribution of the means. Of course, if the parent population is normal, the distribution of the means is normal for all sample sizes.

When the variance of the sampling distribution of the means is to be estimated from sample data, which is usually the case, it will be calculated in the following manner:

$$S_M^2 = \frac{S^2}{N} = \frac{SS}{N(N-1)}$$

The standard error of the sampling distribution of the means is the square root of the variance, of course. The reader's attention is directed to the fact that, for a given standard deviation of the scores, the standard error of the mean is a function of sample size. For example, if the standard deviation of the scores is 12 and the sample size 4, the standard error of the mean is 6. Increasing the sample size to 9, with the same standard deviation of the scores, yields a standard error of the mean equal to 4. The student should complete these calculations for himself to verify the relationship. This is an important idea in sampling statistics—that larger samples yield smaller errors.

19.4 THE SAMPLING DISTRIBUTION OF THE DIFFERENCE BETWEEN THE MEANS OF TWO INDEPENDENT SAMPLES

The behavioral scientist is much more likely to be interested in comparing statistics from two independent experimental samples than comparing a single statistic to a hypothesized parameter as described in the previous section. The usual hypothesis is that there is no difference between the two population means; this is a very common *null* hypothesis.

The theoretical sampling distribution of the difference between the means of two independent random samples from normally distributed populations is likewise normal. The mean of this distribution (the expected value) is equal to the difference between the two population means—if the two parameters are equal, the mean of the distribution is zero. The variance of this distribution is equal to the sum of the variances of the two populations:

$$\sigma^2_{M_1-M_2} = \sigma^2_{M_1} + \sigma^2_{M_2} = \frac{\sigma^2_1}{n_1} + \frac{\sigma^2_2}{n_2}$$

If we assume that the two population variances are equal and substitute σ^2 for σ^2_1 and σ^2_2, the equation becomes:

$$\sigma^2_{M_1-M_2} = \sigma^2 \left(\frac{1}{n_1} + \frac{1}{n_2} \right)$$

When the variance of the sampling distribution of the difference between the means is to be estimated from sample data (which is usually the case), it will be calculated by:

$$S^2_{M_1-M_2} = \frac{SS_1 + SS_2}{n_1 + n_2 - 2} \left(\frac{1}{n_1} + \frac{1}{n_2} \right)$$

This equation makes use of information from both samples to estimate the common population variance. The procedure is known as *pooling the variances.*

The standard error of the sampling distribution is, of course, the square root of the variance:

$$S_{M_1-M_2} = \sqrt{\frac{SS_1 + SS_2}{n_1 + n_2 - 2}\left(\frac{1}{n_1} + \frac{1}{n_2}\right)}$$

19.5 THE SAMPLING DISTRIBUTION OF THE DIFFERENCE BETWEEN THE MEANS OF TWO RELATED SAMPLES

When the behavioral scientist wishes to compare two means from different sets of measures on one group of subjects or from measures on two groups made up of matched pairs of subjects, he must take into account that the two sets of measures are correlated.

If the original measures are normally distributed, the difference between the two sets of measures (let $D_i = X_1 - X_2$ for any pair of measures) and the sampling distribution of the difference between the means ($\bar{D} = M_1 - M_2$) are also normal. The mean of this distribution (the expected value) is equal to the difference between the two population means; if the two parameters are equal, the mean of the distribution is zero. The variance of this distribution is:

$$\sigma_{\bar{D}}^2 = \sigma_{M_1}^2 + \sigma_{M_2}^2 - 2r_{12}\sigma_{M_1}\sigma_{M_2}$$

When this variance must be estimated from sample data, the equation becomes:

$$S_{\bar{D}}^2 = S_{M_1}^2 + S_{M_2}^2 - 2r_{12}S_{M_1}S_{M_2}$$

As a general rule, some computational effort will be saved by approaching the problem in this fashion:

$$S_{\bar{D}}^2 = \frac{SS_D}{N(N-1)} \quad \text{where } SS_D = \Sigma d^2 = \Sigma D^2 - \frac{(\Sigma D)^2}{N}$$

As before, the standard error of the sampling distribution is the square root of the variance:

$$S_{\bar{D}} = \sqrt{\frac{SS_D}{N(N-1)}}$$

Generally, the standard error for two related samples will be somewhat smaller than that for two independent samples.

19.6 NORMAL SAMPLING DISTRIBUTIONS AND PROBABILITY

Given a collection of normally distributed scores with mean μ and standard deviation σ, the student has learned to transform these into z-scores, which may be referred to in Table 3 of the Appendix to determine what proportion of the scores in the collection are above or below any particular score. The equation for this standard score transformation was given as:

$$z = \frac{X - \mu}{\sigma}$$

The student is reminded that the proportion of the area under the normal curve which is to the left of a particular score is the relative frequency of scores in the collection which are lower than this score. Similarly, the proportion of the area which is to the right of a particular score is the relative frequency of scores in the collection which are higher than this score. The student is also reminded that we have defined probability in terms of relative frequency. Let us suppose that the score in the equation above is randomly selected from a population of normally distributed scores with mean μ and standard deviation σ. The probability that the obtained z-score will be equal to or smaller than -1.96 is .025; the probability that the obtained z-score will be equal to or larger than $+1.96$ is also .025; and the probability that the obtained z-score will be equal to or smaller than -1.96 or equal to or larger than $+1.96$ is .05. The student should confirm these by referring to Table 3.

It has been established that the mean of a binomial distribution is np, and the standard deviation is $\sqrt{npq}$. Thus, in the normal approximation of the binomial, a z-score was calculated from:

$$z = \frac{r - np}{\sqrt{npq}}$$

Under the appropriate circumstances (outlined in Section 18.5) a z-score calculated in this manner may be referred to Table 3, and the probability of a particular value of r (or of a higher value or of a lower value) may be determined.

The important idea here is that whenever the normal probability distribution is the appropriate model (and it is for all of the sampling distributions considered in this chapter), and the mean and standard deviation of the sampling distribution are known, a z-score may be calculated from:

$$z = \frac{\text{random variable} - \text{expected value}}{\text{standard error}}$$

These relationships form the basis for the techniques of statistical inference that are described in the following chapters.

19.7 INTERVAL ESTIMATION

Behavioral scientists are often interested in estimating population parameters from sample statistics in such a way that a probability statement can be made about the magnitude of possible error in the estimate. A number of instances of statistical estimation have been considered thus far—the mean and standard error of the mean, for example. Each of these has been a *point estimate*, a single-valued statistic which was determined to be an unbiased estimate of a given parameter. While the possibility of sampling error has been pointed out, nothing has been said with respect to estimating the magnitude of this error.

One of the ways in which the investigator can arrive at a statement about the magnitude of possible sampling error is through the use of *interval estimation*. Assuming that the sampling errors are normally distributed, one may establish a range of values on either side of the point estimate and determine the probability that the parameter lies within this range. This probability is usually expressed as a percentage and is called the *level of confidence*. The probability that the parameter lies within the confidence interval is represented by the lowercase Greek letter gamma (γ).

The reader's attention is directed to Fig. 19.1, which shows the sampling distribution of the means. Notice that 95.46 percent of the total area under the curve lies within plus or minus two standard errors of the mean, and that 4.54 percent of the total lies in the extreme areas beyond this interval. Thus, if the point estimate of the mean were 30 and the standard error of the mean 4, one could be 95.46 percent confident that the population mean lies within the range 22 to 38. The more common practice is to establish the 90 percent, 95 percent, or 99 percent confidence interval.

By referring to a table of areas under the normal curve, one may readily determine that the 95 percent confidence interval (which extends .4750 of the area on either side of the mean) corresponds to plus or minus

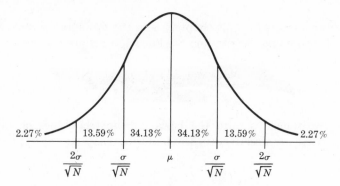

Figure 19.1   Sampling Distribution of the Means.

1.96 standard errors. Formulas for determining the limits of the confidence interval of the mean are simple modifications of the z-score formula:

$$\text{upper limit} = M + z_c S_M$$
$$\text{lower limit} = M - z_c S_M$$

where z_c = 1.645 for the 90 percent confidence interval
1.960 for the 95 percent confidence interval
2.576 for the 99 percent confidence interval

Now, let us illustrate the use of these formulas with data from a sample of size 25 which yields unbiased estimates of the mean and standard deviation, which are 100 and 15, respectively. The standard error of the mean, then, is 3 (15 divided by the square root of 25). Substituting into the formula, we find that the 90 percent confidence interval is 95.0–105.0; the 95 percent confidence interval is 94.1–105.9; and the 99 percent confidence interval is 92.3–107.7. It is apparent that the size of the interval is a function of the level of confidence required, higher levels of confidence resulting in wider intervals. Of course, a large interval does not provide a very precise estimate. The size of the interval is also a function of sample size. Suppose the sample under discussion were increased to 225, and that the estimates of the mean and standard deviation were the same. The standard error of the mean becomes 1, and the 90 percent confidence interval is reduced to 98.4–101.6, the 95 percent confidence interval to 98.0–102.0, and the 99 percent confidence interval to 97.4–102.6. These examples illustrate the very important idea that more precise estimates of population parameters are achieved with larger samples. The use of the normal curve as a model for interval estimation is not recommended for samples smaller than 30 measures. The t-statistic will be substituted for the normal in a later chapter and will permit working with samples of any size.

EXERCISES

19.1   Given a random sample of 64 subjects with sample mean of 50 and unbiased estimate of the population variance equal to 100: $S = 10$
   (a)   Calculate the unbiased estimate of the standard error of the mean.
   (b)   Determine the limits of a 90 percent confidence interval for the mean.

19.2   Given a random sample of 30 subjects with sample mean of 22 and sum of squares equal to 261:
   (a)   Calculate the unbiased estimate of the population standard deviation.
   (b)   Calculate the unbiased estimate of the standard error of the mean.
   (c)   Determine the limits of a 95 percent confidence interval for the mean.

$$19.2 \quad a, \quad \sigma = \sqrt{\frac{\Sigma x^2}{N-1}}$$

$$= \sqrt{\frac{261}{29}}$$

$$= \sqrt{9}$$

$$\sigma = 3$$

$$b. = \frac{S}{\sqrt{N}}$$

$$= \frac{3}{\sqrt{30}}$$

$$= .55$$

$$c, \quad 22 \pm .55 \,(1.96)$$

$$19.1$$

$$S_M = \frac{10}{\sqrt{64}}$$

$$S_m = 1.25$$

$$b. \quad 50 \pm 1.25 \,(1.645)$$

STATISTICAL
HYPOTHESES

20.1 NULL HYPOTHESES

In order to expand the frontiers of man's knowledge, the scientist must b
able to extend with confidence the findings generated by his research fa
beyond the restricted setting in which most research takes place. Th
problem is essentially one of drawing conclusions about some specifi
population from experience which is limited to a relatively small sampl
chosen to represent that population. Arriving at conclusions of this sor
is a necessary hazard with which modern men of science have learned t
live, and it has neither dampened their enthusiasm nor slowed thei
progress in the expansion of knowledge. With the development of th
techniques of statistical inference in the twentieth century, the general
ization of research findings from sample to population may be regarde
as a calculated risk, with the degree of risk defined in terms of the mathe
matics of probability.

The most fruitful approach to the testing of statistical hypotheses ha
been that in which the possible outcomes of an experiment are expresse
as a *null hypothesis* and its alternatives. The null hypothesis is a mathe
matical statement to the effect that there is no difference between specifie
population parameters. Ordinarily, it is a statement to the effect tha
there is no relationship between the independent and dependent variable
in the *population* under study. Of course, this is not usually the anticipate
outcome of the experiment; typically, the investigator conducts th
experiment because he has reason to believe that manipulation of th
independent variable will influence the dependent variable. In othe
words, it is *rejection* of the null hypothesis which is interpreted as a signifi

cant finding. It is important to note that the null hypothesis is a statement about population *parameters;* the hypothesis will be rejected or retained on the basis of sample statistics and a knowledge of the sampling distribution of these statistics.

A hypothesis is rejected if the probability that it is a true statement is very low—lower than some predetermined probability, perhaps one chance in a hundred. This predetermined probability is selected by the investigator, before he collects his data, on the basis of his research experience, the consequences of an incorrect conclusion, and the kind of risk he is prepared to take. This probability is called the *level of significance;* it is represented by the lowercase Greek letter alpha (α), and it is the probability of rejecting a null hypothesis which is in fact true. In other words, it is the probability that the investigator will arrive at a certain kind of erroneous conclusion—the conclusion that a relationship exists between the independent and dependent variables when in fact no such relationship exists in the population under study.

Suppose, for example, that the investigator is interested in studying the relationship between IQ score and academic achievement in American History in the high schools of a very large school district. The scores on both variables for over a million students have been recorded in the central office of the school district. A random sample of 30 pairs of scores is drawn from these central files, and the correlation between the two measures is determined to be .50. The null hypothesis in this case is that the correlation between the two measures is zero in the population under study. To test this hypothesis, the investigator must determine whether a correlation of .50 is sufficiently larger than zero that it is unreasonable to conclude that this correlation is due to sampling error. The sampling distribution of the correlation coefficient is known, and whether this observed coefficient should be attributed to chance may be readily determined, provided that the investigator is prepared to stipulate how big a chance he is prepared to take. Suppose the investigator regards one chance in a hundred as a reasonable risk to take, and from a knowledge of the sampling distribution of the correlation coefficient he determines that there is only one chance in a hundred that a correlation coefficient of .463 or larger will be produced by a sample of size 30 drawn randomly from a very large population in which the correlation is zero. Because the observed correlation coefficient is larger than .463, the null hypothesis is rejected, and the investigator concludes that there is a relationship between the two variables.

Up to this point, the author has used the term *significant* without specifying precisely what is meant. The term is a technical one that statisticians use in a very precise way. When the null hypothesis is rejected, the finding is said to be statistically significant at the specified level of significance. In the example quoted in the previous paragraph, the finding was significant at the .01 level. Rejection of the null hypothesis implies acceptance of some alternative hypothesis. In the example, the investigator concluded simply that there was a relationship between the parameters—that the correlation between the two parameters was larger than zero. No conclusion was arrived at with respect to the magnitude of the relationship between the parameters. However, the correlation coefficient is an unbiased statistic, and the investigator would probably regard the sample statistic as an acceptable estimate of the corresponding population parameter. Of course, he would have greater confidence in the estimate had the statistic been calculated from a larger sample.

When the available evidence does not merit rejection of the null hypothesis, the finding is *nonsignificant*, and the null hypothesis is *retained*. The fact that the investigator did not gather evidence sufficient to reject the null hypothesis does not establish that it is true. At best, when the null hypothesis is retained, it is said to be *tenable*, and the investigator may find it convenient to regard it as true until proved otherwise. As a general rule, the retention of a statistical hypothesis is not a very conclusive finding, and it is very difficult to build a strong argument from a nonsignificant finding.

20.2   THE REGION OF REJECTION

The region of rejection is a proportion of the area in the theoretical sampling distribution which is equal to the level of significance. Generally, this region will be located in one or both tails of the sampling distribution. It represents those sample values that are highly improbable if the null hypothesis is true. The remainder of the area in the theoretical sampling distribution is sometimes referred to as the region of retention. Thus, if a hypothesis is to be tested at the .01 level, 1 percent of the area will be in the region of rejection, and 99 percent will be in the region of retention. Some authors refer to the region of retention as the region of acceptance. This is a poor choice of words, since retention of the null hypothesis should not be equated with acceptance of the null hypothesis.

Table 20.1 is a table of normal areas and z-scores which are presented in a form convenient for hypothesis testing. The information in this table is essentially the same as corresponding parts of Table 3 in the Appendix, except for the manner in which the information is presented. Notice that a distinction is drawn between a *one-tailed* hypothesis test and a *two-tailed* hypothesis test. In the one-tailed test, the area corresponding to the level

Table 20.1   The Normal Model for Hypothesis Testing

Level of significance for a one-tailed test					
.10	.05	.025	.01	.005	.0005
Level of significance for a two-tailed test					
.20	.10	.05	.02	.01	.001
z-scores: 1.282	1.645	1.960	2.326	2.576	3.291

of significance (the region of rejection) is located at one end of the sampling distribution. If the investigator locates this area in the left tail of the distribution, a z-score equal to or smaller than -1.645 is significant at the .05 level. If this area is located in the right tail, a z-score equal to or greater than $+1.645$ is significant at the .05 level. The student should refer to Table 3 in the Appendix to verify that these z-scores correspond to the 5th and 95th percentiles, respectively. In the two-tailed test, the region of rejection is divided into two equal parts, one at each end of the distribution. In this situation, a z-score equal to or smaller than -1.960 or a z-score equal to or larger than $+1.960$ is significant. Again, the student should refer to Table 3 to verify that these z-scores correspond to percentile ranks of 2.5 and 97.5, respectively.

In some research situations, the investigator will have a choice between a one-tailed or a two-tailed test, and his selection will be determined by the manner in which he views his hypothesis and its alternatives. In other situations, the nature of the hypothesis will dictate which type of test is to be used. Generally, when a one-tailed test is the logical choice, it is more likely to produce a significant finding than the two-tailed test. Three examples will be cited to demonstrate how the choice between a one- and two-tailed test is made:

(1) When the significance of a correlation coefficient is tested, the null hypothesis states that the correlation between the two variables is zero in the population under study. There are two alternatives to this null hypothesis: the correlation may be either positive or negative. If the investigator is interested in both alternatives, a two-tailed test is appropriate. If, however, the investigator is prepared to specify that only a positive correlation, for example, will be considered significant, a one-tailed test is appropriate. In this situation, the entire region of rejection

will be located at the positive end of the sampling distribution, and any negative correlation coefficient will be regarded as nonsignificant.

(2) Suppose the investigator is interested in comparing two different methods of instruction. The usual dependent variable will be an examination given at the end of the experiment, and the typical null hypothesis states that there is no difference in the means of the two populations. Again, there are two alternatives to the null hypothesis: Group 1 may have the higher mean, or Group 2 may have the higher mean. Generally, a two-tailed test will be used in this situation. However, if the investigator is prepared to specify in advance that only one of these two alternatives is of interest and will be regarded as significant, he may specify which alternative and choose a one-tailed test.

(3) Consider the situation in which the investigator believes that by showing a certain motion picture to a group of subjects he can cause a positive shift in their attitudes toward members of other races. An attitude scale will be administered before and after the film showing, and the null hypothesis states that there is no difference in the mean scores of the population at the two administrations. Notice that the investigator has stated his interest in a positive shift in attitudes. Only a positive shift in the means will be regarded as significant (a negative shift will be regarded as nonsignificant), and the logical choice is a one-tailed test. Of course, if the investigator had been interested in the possibility of a shift in either direction, a two-tailed test would have been chosen.

When the investigator describes the region of rejection for a statistical test, he will need to indicate the level of significance and whether a one- or two-tailed test is used. The manner in which the region of rejection is described will also be influenced by the type of statistical analysis used. When the normal distribution is used as the theoretical model, the region of rejection will usually be described in terms of z-scores of a certain magnitude. A two-tailed region of rejection at the .02 level would be stated as follows:

$$R: z \leqslant -2.326, \quad z \geqslant +2.326$$

The reader's attention is directed to the ESP experiment reported in Section 18.4, and especially to Table 18.3 with the binomial model for this experiment. The null hypothesis for this experiment states that the number of correct responses given by the subject does not differ significantly from the number of correct responses expected when the responses are completely random. The one-tailed test is appropriate, since the investigator is interested only in rejecting the hypothesis if the subject gives an improbable number of correct answers; he is not interested in rejecting the hypothesis if the subject gives an improbable number of *in*correct answers. Suppose the level of significance were set at .05; the region of rejection may be stated in terms of the number of correct answers.

$$R: r \geqslant 5$$

Notice that the region of rejection in this case is actually about .017; due to the crudeness of the experiment, this is as close to the .05 level as the investigator can get.

20.3 OUTLINE PROCEDURE FOR TESTING STATISTICAL HYPOTHESES

A five-step procedure which is applicable to the testing of a wide variety of statistical hypotheses will be outlined here. Its use will be illustrated with a simple research problem using the normal distribution as a model for the sampling distribution of the means.

A sociologist is interested in studying the relationship between a certain kind of cultural deprivation and adult intelligence test scores. He has identified a population of adult residents of the United States which meets his criterion of cultural deprivation, and he hypothesizes that this population will score below the average on a standardized adult intelligence test. The independent variable is whether or not the individual is a member of the culturally deprived population, and the dependent variable is IQ score as measured by the Wechsler Adult Intelligence Scale (abbreviated WAIS). The WAIS is an individual test of intelligence (it is administered by a professional psychometrist to one individual at a time) which has been standardized on several thousand adults representative of the general populace of the United States. The instrument yields normally distributed scores with mean of 100 and standard deviation of 15 when used with typical adults similar to those in the norm group. The research will be conducted with 100 individuals selected at random from the population of interest.

Step 1 State the null hypothesis and its alternatives in terms of population parameters. It is probably good practice to make both a narrative and a mathematical statement of the null hypothesis. The alternatives may be stated in mathematical form.

In the example, the null hypothesis is that there is no difference in the mean for the culturally deprived population (μ_1) and the mean for the norm group (μ_0). The investigator hypothesizes that the sample will score significantly lower than the norm group, so a one-tailed alternative appears reasonable. The null hypothesis and its alternative may be stated mathematically as follows:

$$H: \mu_1 = \mu_0 \qquad A: \mu_1 < \mu_0$$

Step 2 Choose the appropriate statistic and specify the sample size. The student is not yet acquainted with the large variety of statistical tests available to fit the many different problems encountered in behavioral research. For this particular example, the normal probability distribution will be used as a model for the sampling distribution of the means. The sample size has been set at 100. In practical research, the size of the sample is largely determined by economic considerations and the availability of subjects. The fact that the WAIS is an individual test of intelligence is a limiting factor with respect to sample size in this experiment.

Step 3 Specify the level of significance and define the region of rejection.

Suppose, that for whatever reasons he might have, the investigator chooses to test the hypothesis at the .005 level of significance. The level of significance and the region of rejection are reported as:

$$\alpha = .005 \qquad R: z \leqslant -2.576$$

Step 4 Collect the data and complete the calculations. At this point, the investigator is dealing with sample information.

A mean of 94.3 is calculated from the 100 WAIS scores. The sample mean will now be compared to the sampling distribution of the means. The mean of this distribution is μ_0, which is 100. The standard error of this distribution is $\sigma/\sqrt{N}$, which is 1.5. A z-score may be calculated by taking the deviation of the sample mean from the theoretical mean and dividing by the standard error of the means:

$$z = \frac{M - \mu_0}{\sigma_M} = \frac{94.3 - 100}{1.5} = -3.80$$

Step 5 Complete the test of the hypothesis and interpret the result. If the sample statistic falls within the region of rejection, the null hypothesis is rejected, and the alternative is accepted. If the sample statistic falls within the region of retention, the null hypothesis is retained.

The calculated statistic in this case falls in the region of rejection specified earlier, and the null hypothesis is rejected at the .005 level of significance. Of course, this does not mean that every member of the culturally deprived population has a lower than average IQ, but it does establish, within the limits specified by the level of significance, that the mean for this population is lower than that of the general populace used in norming the instrument.

EXERCISES

20.1 A school psychologist in a Southwestern community is interested in the academic problems encountered in the public schools by children of Spanish-American heritage. He draws a random sample of 64 children with Spanish surnames from the district files and determines that they averaged 47.75 on an English test which had been standardized with mean of 50 and standard deviation of 10. Use the five-step procedure outlined in the text to determine whether these youngsters are significantly lower than the norm group. Use a one-tailed test at the .01 level.

20.2 Refer to the ESP experiment in Section 18.4. Suppose the number of trials were increased to ten, and the subject got eight correct. Write a narrative statement of the null hypothesis and use the five-step procedure outlined in the text to test the hypothesis at the .005 level.

1. $H_0: M_1 = M_2$ $A: M_1 < M_2$ 3 - test $\alpha = .01$ $R: 3 \leq 2.3$

$S_M = \frac{10}{\sqrt{64}}$ $3 = \frac{47.75 - 50}{1.25}$

$S_m = 1.25$ $= \frac{-2.25}{1.25}$

   $3 = -1.8$ H_0 retained

2. $p = 45 \left(\frac{1}{4}\right)^8 \left(\frac{1}{4}\right)^2$

   $= .0034$ less than .005   retain H_0

THE PROBLEM OF
ERROR IN
HYPOTHESIS TESTING

21.1 SAMPLING ERROR

Conclusions drawn about populations from sample information will always be subject to some error as each sample drawn from a given population might be expected to differ somewhat from the parent population. Sampling error is simply the difference between the value of a sample statistic and the corresponding population parameter. While a variety of procedures have been developed for the purpose of minimizing sampling error (matching samples on one or more variables, for example), no procedure available completely eliminates sampling error. The researcher is continually confronted with the possibility that the outcome of his research is attributable to sampling error rather than any real relationship between the dependent and independent variables.

Two kinds of errors are encountered in testing the null hypothesis: (1) rejecting a true hypothesis, and (2) retaining a false hypothesis. These will be treated in considerable detail in the following sections.

21.2 WHEN THE HYPOTHESIS IS REJECTED

Two possibilities exist when the null hypothesis is rejected: (1) the hypothesis is false, or (2) the hypothesis is true.

Consider the situation in which the null hypothesis is in reality false, and the hypothesis is rejected. The estimation of the population parameters by the sample statistics has been sufficiently accurate to produce a valid test of the hypothesis. There is a real relationship between the independent and dependent variables. This is a desirable situation; the investigator has arrived at a true conclusion by rejecting the null hypothesis.

Consider also the situation in which the null hypothesis is in reality true, but the hypothesis is rejected. The estimation of the population parameters by the sample statistics has been sufficiently inaccurate to produce an invalid test of the hypothesis. The investigator concludes that there is a relationship between the independent and dependent variables when no such relationship exists. This is an undesirable situation; the investigator is in error; he has arrived at a false conclusion by rejecting the null hypothesis.

The rejection of a true hypothesis is known as a Type I error. It can only occur when both of these conditions are met: (1) the hypothesis is rejected, and (2) the hypothesis is true. The probability of a Type I error is equal to the level of significance. It is controlled by the investigator; he may set it as high or as low as he wishes. The selection of the .05 level of significance, for example, indicates that there is one chance in 20 of a Type I error—one chance in 20 that the investigator will reject a true hypothesis.

Typically, when rejection is indicated by a statistical test, it is followed by action on the part of the investigator. Recommendations will be made as a result of the research findings; this is a decision making situation, and changes in established procedures may be brought about.

The consequences of a Type I error may be great or small, depending upon the kind of action that results from the research findings. If a Type I error leads to further research, for example, it is likely that the subsequent research will eventually culminate in nonsignificant findings. How serious this is depends upon the amount of time, energy, and other resources expended in the subsequent research.

If a Type I error leads to changes in an educational program, for example, it is unlikely that the changes will prove either beneficial or detrimental to the educational program. How serious this is depends upon the amount of time, energy, and other resources expended in making the changes in the educational program.

Obviously, if important and expensive decisions are to be made from significant research findings, then a rigorous level of significance is needed. The probability of a Type I error must be set very small.

21.3 WHEN THE HYPOTHESIS IS RETAINED

Two possibilities exist when the null hypothesis is retained: (1) the hypothesis is false, or (2) the hypothesis is true.

Consider the situation in which the null hypothesis is in reality true, and the hypothesis is retained. The estimation of the population parameters by the sample statistics has been sufficiently accurate to produce a valid test of the hypothesis (assuming the investigator has collected sufficient evidence to test the hypothesis adequately). There is no real relationship between the independent and dependent variables. This is a desirable situation from a statistical point of view; the investigator has arrived at a true conclusion by retaining the null hypothesis.

Consider also the situation in which the null hypothesis is in reality false, but the hypothesis is retained. Either the estimation of the population parameters by the sample statistics has been sufficiently inaccurate to produce an invalid test, or the investigator has not collected enough evidence to test the hypothesis adequately. There is a real relationship between the independent and dependent variables, but the investigator has failed to establish its existence. This is an undesirable situation; the investigator is in error; he has arrived at an erroneous conclusion by retaining the null hypothesis.

The retention of a false hypothesis is known as a Type II error. It can only occur when both of these conditions are met: (1) the hypothesis is retained, and (2) the hypothesis is false. The probability of a Type II error is sometimes represented by the lowercase Greek letter beta (β). The value of α is established by the investigator, but β is a hypothetical construct, and its value is seldom known in the practical research situation. However, from the study of hypothetical situations in which the population parameters are known, we have established that Type II errors are most likely to occur when the sample size is too small, or when the value of α is set too rigorously. The more the investigator guards against Type I error, the greater the likelihood of Type II error. The value of α should be chosen to minimize Type II error as well as Type I error.

Typically, when retention is indicated by a statistical test, it is followed by inaction on the part of the investigator. No recommendations will be made as a result of the research findings; this is not a decision making situation, and no changes in established procedures are likely to be brought about.

The consequences of a Type II error may be great or small, depending upon what truth is lost as a result of the erroneous nonsignificant finding.

If a Type II error results in the loss of an important idea—an idea which might eventually be of benefit to mankind—the loss could be very great. The investigator who is working on the frontiers of human knowledge, working with new ideas having only vague referents in practical reality, may be more concerned about Type II errors than Type I errors. He may choose to test his hypotheses at the .10 or even the .20 level of significance, for he knows that subsequent research will eventually reveal Type I errors made in the primitive stages of research. No such safeguard exists with respect to Type II errors, however, since these generally discourage further research.

Consider the effects of a Type II error upon research which, for example, is intended to have immediate impact upon educational practices. The investigator may be reasonably sure that the loss is not great, for it is unlikely that research of great practical significance will lead to a statistically nonsignificant finding. This last statement refers only to the situation in which the hypothesis is false; a nonsignificant finding when the hypothesis is true may be of great practical significance.

21.4 CHOOSING THE LEVEL OF SIGNIFICANCE

The level of significance can be intelligently chosen only by weighing the consequences of the two kinds of error. The probability of Type I error is readily controlled by the investigator and is determined by his selection of the level of significance. However, the probability of Type II error is unknown, and, when all other things are equal, the smaller the probability of a Type I error, the greater the probability of a Type II error.

Most behavioral research is conducted at the .01 and .05 levels of significance. However, in exploratory research, the .10 and .20 levels may be more appropriate. When the research leads to the implementation of programs of great cost, it is necessary to guard more rigorously against the Type I error, and the .001 level may be chosen.

Consider the situation in which the investigator may be testing many hypotheses. If 100 hypotheses were tested at the .05 level, the rejection of five true hypotheses should be anticipated. For this reason, when many hypotheses are tested in a given piece of research, it is probably desirable to choose a more rigorous level of significance.

The investigator may reduce the probability of a Type II error by increasing the size of his sample or by using a more powerful statistical test. Sample size and the power of a statistical test are considered in the following sections.

21.5 SAMPLE SIZE

Sampling error has been defined as the difference between the value of a sample statistic and the corresponding population parameter. It should be immediately apparent that sampling error is a function of sample size, and that the error tends to be smaller for larger samples. This relationship was illustrated in Section 19.7. The formulas used for determining the limits of confidence intervals may be modified slightly to provide a formula for sample size, provided that the investigator is prepared to specify how much error is acceptable and how much confidence is required. Let e_m be the maximum acceptable error and z_c the z-score corresponding to the required level of confidence. Then, for the sampling distribution of the means:

$$e_m = z_c \sigma_M \quad \text{where } \sigma_M = \frac{\sigma}{\sqrt{N}}$$

With a little algebra, one may demonstrate that:

$$N = \left(\frac{z_c \sigma}{e_m}\right)^2$$

For example, suppose the investigator desires to obtain an estimate of the mean IQ (where $\sigma = 15$) for a population of interest, and that he wishes to be 95 percent confident that the population mean is within three points of the sample mean. The required sample size is:

$$N = \left(\frac{1.96 \times 15}{3}\right)^2 = 96$$

This calculation is, of course, dependent upon a knowledge of the standard deviation of the population under study. This information is not usually available and must be estimated from sample information. Typically, the investigator is not prepared to estimate the standard deviation from sample data until after the research is completed. This is no problem if he is willing to state the maximum acceptable error as a fraction of the standard deviation. Suppose, for instance, the investigator specifies that e_m is to be $\sigma/10$ at the 95 percent level of confidence; the calculation becomes:

$$N = (1.96 \times 10)^2 = 384$$

These calculations are based upon the assumption that the investigator is sampling from a population of infinite size—a common assumption in statistical inference.

The author prefers to approach the problem of sample size with some rules of thumb which he believes are appropriate for determining sample size needed in most behavioral research. Generally, samples smaller than 30 in size are to be avoided, especially with the parametric statistics. The use of samples of size 30 or larger usually insures for the investigator the benefits of the central limit theorem (see Section 19.3 for a discussion of this theorem). There are few occasions in behavioral research in which the use of samples larger than 500 in size is merited. A sample of size 500 from a population of infinite size assures that the sampling error will not exceed $\sigma/10$ about 98 percent of the time. Within these limits—from about 30 to 500—the use of a sample about one-tenth as large as the parent population is recommended. If the sample is to be broken into subsamples—suppose, for example, that males and females are to be analyzed separately—the recommendations with respect to minimum sample size apply to each of the subsamples.

When, out of necessity, the investigator must work with smaller samples, the use of matching techniques or similar procedures to increase the precision of the experiment is recommended.

21.6 THE POWER OF A STATISTICAL TEST

The *power* of a statistical test of some hypothesis is defined as the probability that it will reject the hypothesis when the hypothesis is false. The power of a test plus the probability of a Type II error is equal to one. Typically, neither is known in the practical research setting. However, these are useful constructs that enable the investigator to make the most intelligent use of the techniques of statistical inference.

Generally, the power of a given statistical test is a function of the sample size, of the level of significance used, and of just how false the hypothesis is. These ideas are illustrated for the test of a hypothesis about the mean of a normally distributed population in Fig. 21.1. The graph con-

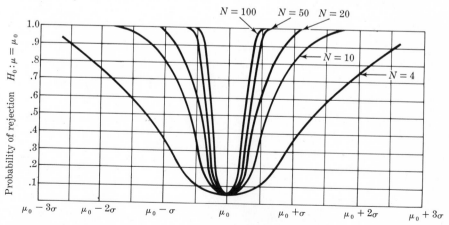

Figure 21.1   Power Curves of a Statistical Test.

sists of a family of power curves, each curve representing a different sample size, for the null hypothesis $\mu = \mu_0$. The level of significance has been set at .05, and a two-tailed test is used. Notice that the probability of rejecting the null hypothesis when the hypothesis is true (the probability of a Type I error) is represented by the convergence of the curves in the center of the graph. Suppose, however, that the population mean is equal to $\mu_0 + \sigma/2$. With a sample of size 10, the probability of rejecting the hypothesis (which is false) is .37, and the probability of retaining the hypothesis (the probability of a Type II error) is .63. When the sample size is increased to 20, the probability of rejecting the hypothesis is .62, and the probability of retaining the hypothesis is .38. Of course, when it is appropriate to use, a one-tailed test is more powerful than a two-tailed test in the same situation.

Some statistical tests are more powerful than others. The t-tests and the F-tests are usually more powerful than other statistical tests of the more common null hypotheses. The *relative power* (or *power-efficiency*, as it is sometimes called) of a statistical test is a measure of its ability to detect a false hypothesis as compared to a standard test, such as the t-test or the F-test. The usual procedure for expressing relative power is to form a ratio of the sample size needed to reject the hypothesis with the standard test to the sample size needed to reject the hypothesis with the test under consideration. Thus, if it should be determined that a given false hypothesis is rejected by the t-test with a sample of size 90, but it requires a sample of size 100 to reject the same hypothesis with the Mann-Whitney U-test, the Mann-Whitney is said to be 90 percent as powerful as the t-test.

THE *t*-DISTRIBUTION

22

22.1 THE DEVELOPMENT OF SMALL-SAMPLE STATISTICS

The early history of statistical inference is that of large-sample statistics using the normal curve as a model for the sampling distribution. While the normal curve provides a probability distribution that is fairly well approximated when drawing large samples, it has proved inappropriate for small samples, especially those smaller than 30 in size. The problem arises from the fact that there are usually two distinct sources of sampling error in the typical test of a practical research hypothesis. The first of these, estimation of the parameters identified in the statement of the null hypothesis, is taken into consideration in the normal model of the sampling distribution. The second source of sampling error, estimation of the standard error of the sampling distribution, is not. When the standard error is estimated from information provided by a small sample, the error may be of considerable consequence, especially in the tails of the sampling distribution where hypotheses are tested.

Shortly after the turn of the century, a young chemist named William S. Gossett was employed by a brewery in Dublin, Ireland. Gossett was concerned with quality control, and he approached the problem by drawing samples and analyzing these with the traditional statistical procedures. He soon recognized the inadequacy of the normal curve as a probability model for small samples, but lacking the mathematical sophisitication needed to derive new models, he determined to establish the necessary sampling distributions empirically.

Gossett secured the records of Bertillon measurements for some 3000 British criminals. The Bertillon system was the first scientific method for criminal identification and was based largely on classification of body measurements. He recorded the measurements on cards, building an approximately normal population from which he could conveniently draw random samples of various sizes. By drawing large numbers of samples of a given size, he was able to establish empirically the sampling distribution of the means for that size sample. He established a different sampling distribution for each sample size of interest to him, and he discovered that the distributions for small samples differed markedly from the normal. Later, with the assistance of a university professor, Gossett was able to derive a general mathematical expression for these sampling distributions, and published the results in 1908 under the pen name, "Student." His findings received little attention until they were included in the first textbook of modern statistics, written almost 20 years later by Sir Ronald A. Fisher. The development of Student's t-distribution was the major breakthrough from classical to modern statistical inference.

22.2 THE STUDENT'S t-DISTRIBUTION

A t may be calculated which closely resembles the z used in testing the hypothesis about the mean in Chap. 20. The numerator of the t will be the deviation of the sample mean from the hypothesized mean, and the denominator will be the unbiased estimate of the standard error of the mean:

$$t = \frac{M - \mu}{s_M} \quad \text{where } s_M = \sqrt{\frac{SS}{N(N-1)}}$$

The equation of the t-distribution is much too complex to be treated here. There is a family of t-curves, each determined by a single parameter, its degrees of freedom, a concept closely related to sample size. Figure 22.1 is a graphic illustration of the t-distribution for one, three, five, and infinite degrees of freedom. The distribution for infinite degrees of freedom is identical to the normal. Table 7 in the Appendix gives the distribution of t for many different degrees of freedom and for those levels of significance commonly used in testing hypotheses.

Like the standard normal curve, the various t-distributions are symmetrical and bell-shaped with mean of zero and unit standard deviation. When the number of degrees of freedom (abbreviated df) is small, the proportion of the area in the tails is much greater than that of the normal distribution. For example, .046 of the area of the normal curve lies beyond plus or minus two standard deviations from the mean, but the corresponding points on the t-curve for three degrees of freedom mark off .140

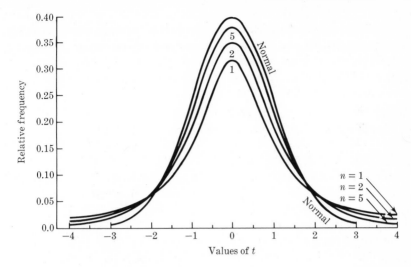

Figure 22.1 The t-Distribution for Various Degrees of Freedom.

of the area under the curve. This suggests that there is greater probability of error when dealing with small samples, and rejection of the null hypothesis is not as easily achieved.

22.3 DEGREES OF FREEDOM

The number of degrees of freedom of a statistic depends upon the number of observations (N) and the nature of the problem. The term "degrees of freedom" is a descriptive one which suggests its general nature, but a full appreciation of the term is not likely to be realized without some very serious study of theoretical statistics. This will not be undertaken in this text. The approach used here shall be:

(1) The term will be defined, even though it is anticipated that the definition will not be particularly meaningful to many students.

(2) For each type of statistical calculation encountered in this text, the number of degrees of freedom will be clearly specified.

(3) The manner in which the definition was used to arrive at the number of degrees of freedom will be explained.

First, the definition: the number of degrees of freedom is equal to the number of observations minus the number of algebraically independent linear restrictions placed upon them. Generally, these restrictions will be identified as statistics which must be calculated from sample data before it is possible to calculate the statistic used in testing the hypothesis. Consider the estimation of the standard error of the mean:

$$s_M = \sqrt{\frac{SS}{N(N-1)}} \qquad \text{where } SS = \Sigma\,(X_i - M)^2$$

Notice that it is necessary to estimate the mean from sample data before the standard error of the mean can be estimated. The price for this procedure is one degree of freedom, and the degrees of freedom for the standard error of the mean is $N-1$.

Another way of looking at degrees of freedom is to think in terms of the number of values of the variable that are free to vary. Suppose, for example, that a sample of size four is drawn, the scores are 2, 3, 5, 6, and the mean is 4. The deviation scores are -2, -1, $+1$, $+2$, and the algebraic sum of the deviation scores is, of course, zero. As a consequence, if any three of the deviation scores were calculated, the value of the fourth is not free to vary; its value must be such that the algebraic sum of the deviation scores is zero. This is equivalent to saying that if the mean is specified, the value of the last score drawn is not free to vary. The degrees of freedom in the example is three, as only three scores are free to vary.

When the sum of the squared deviations about the mean was calculated, only one parameter was estimated from sample data, and the degrees of freedom for the standard error of the mean was determined to be $N-1$. When the sum of the squared deviations about the regression line was calculated, however, both the slope of the line and its intercept with the Y-axis were estimated from sample data, and the degrees of freedom for the standard error of estimate was determined to be $N-2$.

It is customary to arrange the values of t in tabular form for all degrees of freedom from one to 30, for a few larger values, and for infinity. The t-distribution for infinite degrees of freedom is identical to the normal and will be substituted for the normal in hypothesis testing. Where the table is incomplete and does not include the exact degrees of freedom specified for a given problem, the appropriate thing to do is to use the table value closest to the desired value, but on the low side. Discarding degrees of freedom in this fashion gives a conservative approach to hypothesis testing, and rejection will be slightly less likely. This recommendation is not limited to the t-distribution, but will be used with other statistics as well.

22.4 THE TEST OF A HYPOTHESIS ABOUT THE MEAN

The t-distribution may be used as a model for testing a hypothesis about the mean of a normally distributed population, just as the normal distribution was used in Chapter 20, but with the additional advantage that the *t*-test is valid for samples of any size. In order to use this test, the investigator must be able to specify some hypothesized mean to which he wishes to compare the mean of the population under study. Typically, the procedure will be used to determine whether the population of interest differs significantly from the norm group used in standardizing the measuring instrument.

The test assumes that the sampling distribution of the means is normal. This is true for samples of any size drawn from normal populations, and it is satisfactorily approximated by samples of adequate size from non-normal populations. If the departure is moderate, as in the case of the typical examination prepared by a classroom teacher, a sample of 30 measures is probably adequate. If the departure from normality is more than moderate, a larger sample is required. A sample of size 100 is adequate for almost any research situation.

The procedure for conducting the investigation is outlined as follows:

(1)   A random sample of measures is drawn from the population under study, and the mean of this sample (M) is calculated.

(2)   With μ representing the hypothesized mean and s_M representing an unbiased estimate of the standard error of the mean, the t is calculated by:

$$t = \frac{M - \mu}{s_M} \quad \text{where } s_M = \sqrt{\frac{SS}{N(N-1)}}$$

(3)   At some predetermined level of significance, the calculated t is compared to the tabled value with the appropriate number of degrees of freedom ($df = N - 1$). A one- or two-tailed test may be used. If the calculated value equals or exceeds the tabled value, the finding is significant, and the hypothesis is rejected; the difference between the two means is significant at the specified level. If the calculated value is smaller than the tabled value, the hypothesis is retained; there is no significant difference between the means. (See Table 7.)

22.5   INTERVAL ESTIMATION

The *t*-distribution is to be preferred over the normal for constructing confidence intervals for small samples. While confidence intervals may be constructed for all of the common statistics, the confidence interval for the mean is that which is most often desired. The formulas for the upper and lower limits are the same as those used when calculating confidence limits with the normal model, except that the appropriate t is substituted for the z.

As always, two pieces of information—the level of significance and the degrees of freedom—will be needed to enter the t-table. The level of significance, in this case, will be equal to one minus the level of confidence, and the two-tailed t will be used. The degrees of freedom will be the same as that used in calculating the unbiased estimate of the standard error of the mean, and this will usually be $N - 1$.

The formulas for the confidence limits are as follows:

$$\text{upper limit} = M + t_c s_M$$
$$\text{lower limit} = M - t_c s_M$$

The data which served to illustrate the use of the normal model for interval estimation will also be used to illustrate the use of the t for this purpose. A sample of size 25 was drawn, the sample mean was 100, and the standard error of the mean was determined to be 3. For the 90 percent confidence interval with 24 degrees of freedom, the t is 1.711, and the confidence limits of the mean are 94.9–105.1 for the example. This interval is slightly larger than that yielded by the normal model; it would have been much larger had a smaller sample been used.

EXERCISES

22.1 Suppose an investigator wishes to test the hypothesis that the mean of a certain population is 30. A random sample of ten measures is drawn:

$$5, \ 15, \ 20, \ 20, \ 25, \ 25, \ 30, \ 35, \ 35, \ 40$$

Using the five-step procedure for hypothesis testing, test the hypothesis that there is no difference between the population mean and the hypothesized mean. Use the .05 level and a two-tailed test.

22.2 Calculate the 95 percent confidence limits for the mean of Exercise 22.1.

THE *t*-TEST FOR TWO INDEPENDENT SAMPLES

23.1 FUNCTION

The situation in which the researcher assigns one group of subjects to an experimental treatment and uses another group of similar subjects as a control group is one of the most popular experimental designs. This is equivalent to using two experimental treatments, and our discussion will not distinguish between the use of a control group or of a second experimental group. For example, in a learning experiment, the experimental group might be subjected to a new instructional technique, while the control group receives the traditional instruction. The traditional instruction is simply a second experimental treatment. In a medical experiment, one group of subjects might receive a new medicine, while the second group receives a placebo (a medicine with no active ingredients). Whether the investigator chooses to regard the subjects receiving the placebo as a control group or as a second experimental group is of no consequence. The experimental treatments or other variable which distinguishes the two groups from each other is the independent variable.

At the conclusion of the experiment, all subjects in both groups are measured with the same criterion—a psychological test or other instrument intended to determine whether anything of consequence took place as a result of the experiment. The criterion is also called the dependent variable. The *t*-test for two independent samples is used to determine whether the criterion means for the two groups differ significantly.

It is also permissible to use the *t*-test for two independent samples to examine existing differences between random samples from two different populations. No experimental treatments are administered by the investigator in this situation; he is seeking to determine whether the two popula-tions differ significantly on some criterion variable. This is equivalent to testing the null hypothesis that, with respect to the criterion variable, the two populations are in fact a single population. Thus, an investigator might be interested in determining whether a collection of children should be regarded as a single population, or as two populations, one male and the other female. Similarly, with respect to a given criterion, one might be interested in determining whether a given group of psychology students is simply a group of psychology students, or two groups, one consisting of psychology majors and the other consisting of nonpsychology majors.

23.2 RATIONALE

In order to test a hypothesis about the difference between the means of two independent samples, one must first be able to estimate the standard error of the difference between the means of two independent samples. This statistic was identified in the chapter on sampling theory, and it may be estimated from:

$$ s_{M_1 - M_2} = \sqrt{\frac{SS_1 + SS_2}{n_1 + n_2 - 2} \left(\frac{1}{n_1} + \frac{1}{n_2} \right)} $$

The reader should note that there are two important assumptions underlying this estimate: (1) the distribution of the measures in both samples is normal, and (2) the variances of the two populations are equal. Moderate departures from these assumptions have proved to be of no practical consequence. When the two samples are of equal or near equal size, the statistical test is quite insensitive to violations of these assumptions.

Dividing the difference between the two means by the standard error of the difference yields a statistic which is distributed as the t-distribution if the two population means are equal. To the extent that the two means are not equal, we would expect the value of the calculated t to be inflated, and the probability of rejecting the null hypothesis becomes greater than the level of significance.

23.3 PROCEDURE

(1) Two random samples are independently drawn from the population under study, and these are subjected to two different experimental treatments.

(2)   At the conclusion of the experiment, a single criterion measure is administered to all subjects, and the mean of each group is calculated. The standard error of the difference between the means is calculated, using the formula given in the previous section.

(3)   The t is calculated by:

$$t = \frac{M_1 - M_2}{s_{M_1 - M_2}}$$

(4)   The calculated t is compared to the tabled value at the desired level of significance and with degrees of freedom equal to $n_1 + n_2 - 2$. A one- or two-tailed test may be used, although the two-tailed test is more common with this design. If the calculated statistic equals or exceeds the tabled value, the hypothesis is rejected, and the observed difference between the two means is a significant one. The two experimental treatments are determined to produce different results at the specified level of significance, and the treatment which produced the higher mean is usually regarded as superior, although the exact interpretation is dependent upon the nature of the dependent variable. If the calculated value is smaller than the tabled value, the hypothesis is retained, and no significant difference between the means has been established.

23.4   EXAMPLE

A researcher is studying the effects of two different diets on weight gain of white rats. Two random samples of size ten each are chosen from an available colony of adolescent male white rats. One sample is fed Diet A and the other Diet B during the experimental period. The rats are weighed before the experiment and again following the experiment; the criterion variable is gain in weight. The test of the null hypothesis is recorded below, complete with all sample data:

(1) $H: \mu_A = \mu_B \qquad A: \mu_A \neq \mu_B.$

(2) t-test for independent samples, $n_A = n_B = 10.$

(3)   .05 level, two-tailed, $df = 18.$
$R: t \leqslant -2.101, \quad t \geqslant +2.101$

(4)   Sample A: 1, 2, 3, 4, 4, 5, 5, 8, 9, 9

$$n_A = 10, \quad M_A = 5, \quad SS_A = 72.$$

Sample B: 4, 6, 7, 7, 8, 8, 9, 10, 10, 11

$$n_B = 10, \quad M_B = 8, \quad SS_B = 40.$$

$$s_{M_A - M_B} = \sqrt{\frac{72 + 40}{10 + 10 - 2} \left(\frac{1}{10} + \frac{1}{10} \right)} = 1.12.$$

$$t = \frac{5 - 8}{1.12} = -2.67.$$

(5) Reject the null hypothesis. There is a significant difference in the means of the two samples. The data *suggest* that Diet B secures greater weight gains than Diet A.

23.5 DISCUSSION

The t-test for two independent samples is one of the most popular statistical tests. It is a very powerful test that lends itself to a wide variety of research problems.

There are certain assumptions underlying the use of this t-test, and it is important to note that these assumptions apply to the two populations specified in the statement of the null hypothesis. In the experimental setting where the investigator actually manipulates the independent variable, the two populations do not exist in reality—they are hypothetical populations which have been subjected to experimentation, and from which the two samples (after experimentation) could be regarded as random samples. After experimentation, the two samples are no longer random samples from the original parent population. The use of the t-test for two independent samples assumes that the dependent variable is measured on at least an interval scale, that it is normally distributed in the two populations, and that the two populations have equal variances. A final assumption, that the means of the two populations are equal, is necessary for the distribution of the calculated t to match that of the tabled t. When the hypothesis is rejected, we prefer to assume that it is this final assumption which has been violated. Fortunately, this t-test is much more sensitive to violation of the assumption of equal means than to violation of the assumptions of normality or of homogeneous variances. Tests for normality and for equal variances are discussed in other parts of this text (see Sections 28.5, 31.5, and 34.2). Unfortunately, tests for normality are not very effective with small samples, and the benefits of the central limit theorem eliminate the need for such tests with large samples. The most powerful tests for equal variances are very sensitive to departures from normality and are of limited value for testing the assumption of homogeneous variances. Although some other tests are available, they are quite feeble and provide little support for the investigator when the hypothesis of equal variances is retained. Generally, research with respect to the *t*-distribution suggests that when the investigator is working with samples of the same size or nearly the same size, he may ignore these assumptions unless he has reason to believe that his measures deviate greatly from them. The one exception appears to be the situation in which the assumptions of normality and homogeneous variances are both violated, in which case the probability of a Type I error is likely to be somewhat larger (perhaps twice as large) than the level of significance.

The Mann-Whitney *U*-test is a nonparametric alternative to the *t*-test which may be used with ordinal measures or with data that deviate from the normal or in the situation where heterogeneous variances are encountered. It is almost as powerful as the *t*, but it becomes less powerful if many tie scores are encountered. The median test, which uses the chi-square distribution, is another nonparametric alternative, but it is not as powerful as the Mann-Whitney.

EXERCISES

23.1 Air Force psychologists are conducting research into the relative effectiveness of two methods of training navigators. The first method makes use of computer-simulated flight; the second uses traditional classroom instruction. Two random samples of size ten each are selected for participation in the experiment; however, two of the men in the first group get chicken pox and are dropped from the experiment. The criterion score is a performance test which has been scaled to yield a stanine score. The criterion scores for the two samples are as follows:

> *Group I (simulated flight):*   2, 5, 5, 6, 6, 7, 8, 9
> *Group II (traditional):*   1, 2, 2, 3, 4, 4, 5, 5, 6, 8

Use the five-step procedure for testing the null hypothesis at the .05 level.

23.2 Refer to Exercise 12.3. Use the *t*-test for two independent samples to test the hypothesis of no difference between boys and girls at the .05 level. Use a one-tailed alternative that girls will score higher than boys.

23.3 Given the following data:

$$n_1 = 30, \quad M_1 = 48.2, \quad SS_1 = 1740$$
$$n_2 = 29, \quad M_2 = 52.4, \quad SS_2 = 1622$$

test the hypothesis of equal means using a two-tailed test at the .05 level.

THE *t*-TEST FOR TWO RELATED SAMPLES

24.1 FUNCTION

The *t*-test for two independent samples is a simple randomized design; it is an experimental design in which randomization is the only provision for controlling the contribution of irrelevant variables. When some additional provision is made for the control of these variables, the size of the error term is reduced, and a more precise experimental design is achieved.

Under certain circumstances, it is possible to increase greatly the precision of an experiment by switching from two independent samples to two related samples. This calls for the *t*-test for two related samples, and it fits a number of experimental designs.

One way to increase the precision of an experiment is to draw a single random sample and administer both experimental treatments (at different times) to the same group of subjects. The criterion measure must be administered following each treatment. Of course, this procedure is valid only when there is no carry-over from one treatment to the next and from one measurement to the next. This design is not likely to be practical in research where learning takes place; however, it is quite practical in some psychophysical and medical experiments. Whenever this procedure is valid, an extremely precise research design is possible due to the unique control of irrelevant variables. Designs in which the criterion is administered more than once to the same group of subjects are called *repeated measurements* designs.

Occasionally, the experimenter will be able to use an experimental group as its own control. In this case, the criterion measure is administered to the sample before and after the experimental treatment. The test of

170

significance is used to determine whether any change has taken place. Of course, the experimenter must be on the lookout for effects other than the experimental treatment which might cause a significant shift in the criterion variable.

Finally, this type of analysis may be used with an experiment in which two random samples have been matched on one or more variables. Typically, the design calls for identifying a variable which is known to correlate with the criterion (IQ or a pretest on the criterion are common in learning research). The population is then split into pairs on the basis of the pretest measure, each subject being matched as closely as possible with another subject. A random sample of pairs of subjects is drawn, and one member of each pair is randomly assigned to each of the two treatments. This design is especially valuable to the investigator who must work with a very small population—handicapped children, for example. It is not likely to be as precise as a repeated measurements design, but it is applicable to learning research.

24.2 RATIONALE

In order to test a hypothesis about the difference between the means of two related samples, one must first be able to estimate the standard error of the difference between the means of two related samples. This statistic was identified in the chapter on sampling theory, and it may be estimated from:

$$S_{\bar{D}} = \sqrt{\frac{\Sigma\, d^2}{N(N-1)}}$$

where
$$\Sigma\, d^2 = \Sigma\, (D - \bar{D})^2 = \Sigma\, D^2 - \frac{(\Sigma\, D)^2}{N}$$

The symbol D represents the difference between a pair of scores, $\bar{D}$ the mean of these differences or $M_2 - M_1$, and N the number of *pairs* of scores. The degrees of freedom for this error term is $N - 1$ since it is calculated from N differences, and $\bar{D}$ is estimated from sample data.

To use this as the error term in a *t*-test, one must assume that the distribution of the difference scores is normal—a reasonable assumption in practically any behavioral research using this statistic.

24.3 PROCEDURE

(1) The procedure for conducting the experiment is largely determined by the selection of one of the three designs discussed in Section 24.1. Where two experimental treatments are administered to the same group of subjects, a single random sample is drawn, the two treat-

ments administered (at different times) to all subjects, and a single criterion measure is administered twice, once after each treatment. It is usually desirable to randomize the order of the two treatments, with half of the subjects getting one treatment first, and the other half getting the other treatment first. This cannot be expected to compensate completely for any carry-over from treatment to treatment or measure to measure, however, and the design should not be used if carry-over is suspected.

In the situation where the experimental group is used as its own control, a single random sample is drawn, and the criterion is administered both before and after the experimental treatment.

When matched pairs are used with two different experimental treatments, paired random samples are drawn as described in Section 24.1. The conduct of the experiment will be much like that with two independent samples, however, the criterion measures will be paired for purposes of analysis.

(2) At the conclusion of the experiment, the difference between each pair of scores and the difference between the two means are calculated. When a group is used as its own control, the first score should be subtracted from the second (and the first mean from the second), since this yields a positive difference for a gain on the criterion. The standard error is calculated using the formula given in the previous section.

(3) The t is calculated by:

$$t = \frac{\bar{D}}{S_{\bar{D}}} \quad \text{where } \bar{D} = M_2 - M_1 \quad \text{and} \quad D = X_2 - X_1$$

(4) The calculated t is compared to the tabled value at the desired level of significance and with degrees of freedom equal to $N - 1$ (where N is the number of pairs of scores). A one- or two-tailed test may be used. If the calculated statistic equals or exceeds the tabled value, the hypothesis is rejected, and it is concluded that the observed difference between the two means is a significant one. The exact interpretation of a significant finding is dependent upon the experimental design used and the nature of the criterion variable. If the calculated t is smaller than the tabled value, the hypothesis is retained, and no significant difference between the two means has been established.

24.4 EXAMPLE

A researcher is studying the relative merits of two different methods of teaching mentally retarded children to solve simple problems. There are

20 such children available to participate in the research, and these are grouped in ten pairs, with the two children in each pair matched as closely as possible on such variables as age, IQ, social development, etc. The children are randomly assigned to two training classes, with one member of each pair going into each class. One class is taught by Method A and the other by Method B. At the end of the training period, each child is administered an examination consisting of 20 problems, and his score is recorded as the number of problems solved correctly. The data yielded by the experiment are recorded in Table 24.1.

Table 24.1   Data from the Problem Solving Experiment

Pair	Method A	Method B	D	D²
1	12	16	4	16
2	11	9	−2	4
3	12	15	3	9
4	10	10	0	0
5	10	9	−1	1
6	9	12	3	9
7	7	9	2	4
8	6	10	4	16
9	4	9	5	25
10	4	3	−1	1
			$\Sigma D = 17$	$\Sigma D^2 = 85$

The test of the null hypothesis is outlined below:

(1) $\mu_A = \mu_B$ $A: \mu_A \neq \mu_B$.

(2) *t*-test for related samples, $N = 10$.

(3) .05 level, two-tailed test, $df = 9$.
     $R: t \leqslant -2.262, \quad t \geqslant +2.262$.

(4) $\bar{D} = 17/10 = 1.70$.

$$\Sigma d^2 = 85 - \frac{17^2}{10} = 56.1.$$

$$S_{\bar{D}} = \sqrt{\frac{56.1}{10(10 - 1)}} = .79.$$

$$t = \frac{1.70}{.79} = 2.15.$$

(5) Retain the null hypothesis. The difference between the sample means is nonsignificant.

24.5 DISCUSSION

The t-test for related measures is a very powerful statistical test that may be used with certain kinds of experimental designs. Its great advantage over the t-test for independent samples is the increase in precision accomplished by the reduction in sampling error. Its use with two experimental treatments in the repeated measurements design is restricted to the situation in which the treatment effects are known to be temporary. There must be no carry-over from treatment to treatment or from measure to measure. Obviously, this design is of limited value in learning research.

The use of this statistical test assumes that the distribution of the differences between the two sets of criterion measures is normal in the populations specified by the null hypothesis. This is a reasonable assumption in practically any behavioral research and should be of little concern to the investigator unless he has reason to believe otherwise. The use of the test also assumes measurement on at least an interval scale.

Of the available nonparametric alternatives to the t-test for related samples, the Wilcoxon matched-pairs signed-ranks test is probably the best. It is intended for use with ordinal data.

EXERCISES

24.1   An investigator has developed an individualized program of instruction in problem solving which he believes will raise the IQ of an individual. A random sample of eight children is chosen from a junior high school physical education class to participate in the instructional program. Their IQ scores are assessed before and after the instructional program using equivalent forms of the same instrument. The scores are recorded below:

Before: 81 84 90 97 108 111 118 124
After: 89 88 94 96 118 111 121 121

Use the five-step procedure for testing the null hypothesis at the .05 level.

24.2   A sociologist is studying the effects of a certain motion picture film upon the attitudes of Negro students toward white students. He hypothesizes that viewing the film will cause the scores of the students on a certain attitude scale to shift downward. The scores of twelve participating students are recorded below:

Before: 10 13 18 12 9 8 14 12 17 20 11 7
After: 5   9 13 17 4 5 11 14 13 18 10 7

Use the five-step procedure for testing the null hypothesis at the .05 level.

THE MANN-WHITNEY
U-TEST

25.1   FUNCTION

The Mann-Whitney *U*-test is a nonparametric alternative to the *t*-test for two independent samples. It is commonly used where the experimenter draws two random samples from the same parent population, subjects each to a different experimental treatment, and compares the two on a single criterion. Of course, the experimenter may designate one of his samples as a control group, in which case the experimental treatment is no treatment. The *U*-test may also be used in the situation where independent random samples are drawn from two different parent populations and compared on a single criterion to determine whether the two populations differ.

The Mann-Whitney *U*-test requires data on at least an ordinal scale, and this data is assumed to be continuously distributed. It does not require normality of distribution nor homogeneity of variance for the groups under study. This is one of the most useful of the nonparametric tests; it is almost as powerful as the *t*-test under common research conditions, and the computations are quite simple. It is especially useful with small samples, since violations of the assumptions underlying the *t*-test are most likely to be detrimental and most likely to go undetected with small samples.

25.2   RATIONALE

Suppose a measure on some specified continuous variable is available for every member of a population of interest. Let us draw two independent random samples from this population, and let n_a be the size of Sample A, n_b the size of Sample B, and N the sum of n_a plus n_b.

Now let us combine the two sets of scores and rank them from one to N, using the subscripts a and b to identify each rank as belonging to a particular sample. If the original scores were truly continuously distributed (no tie scores), the number of possible combinations of ranks is equal to $\binom{N}{n_a}$ or $\binom{N}{n_b}$, these being mathematically equivalent. If the two samples were indeed independent random samples from the same parent population, each of these combinations is equally probable. If, however, the samples came from two populations which differ on the criterion variable, there will be a tendency for the ranks of Sample A to cluster at one end of the distribution and those of Sample B to cluster at the other end.

Let U_a be the number of times a rank from Sample A is preceded by a rank from Sample B, and U_b the number of times a rank from Sample B is preceded by a rank from Sample A. For example, consider this set of data.

$$1_a, \ 2_a, \ 3_b, \ 4_a, \ 5_b, \ 6_b$$
$$U_a = 0 + 0 + 1 = 1$$
$$U_b = 2 + 3 + 3 = 8$$

Let us illustrate the calculations by concentrating on U_b: the first b-score (3_b) is preceded by two a-scores, the second b-score (5_b) by three a-scores, and the third b-score (6_b) by three a-scores. Thus, the value of U_b is 8. With a little algebra, one may demonstrate that:

$$U_a = n_a n_b - U_b \quad \text{and} \quad U_b = n_a n_b - U_a$$

We will concentrate our attention on the smaller of these two statistics, and designate it as U.

The value of U will be zero if all of the ranks from Sample A are clustered at one end of the distribution and all of the ranks for Sample B at the other. That is, U equal to zero represents the greatest difference possible between the two samples. The value of U will increase as the ranks are intermixed, that is, as the samples become more alike.

Refer to the data given earlier; the number of combinations for N equal to six and n_a equal to three is 20. The probability that all of the ranks of Sample B would be larger than the ranks of Sample A is 1/20. The value of U in this case would be zero. Thus, U equal to zero would be the region of rejection for a one-tailed test at the .05 level or for a two-tailed test at the .10 level. In similar fashion, one could establish the probabilities associated with different values of U (values less extreme than zero) and for different size samples. The values of U needed to reject the null hypothesis at various levels of significance and for various sample sizes will be found in Table 8 of the Appendix.

It will, as a general rule, be more convenient to determine the value of U from these formulas:

$$U_a = n_a n_b + \frac{n_b(n_b + 1)}{2} - \Sigma R_b$$

$$U_b = n_a n_b + \frac{n_a(n_a + 1)}{2} - \Sigma R_a$$

where ΣR_a is the sum of the ranks for Sample A, and
ΣR_b is the sum of ranks for Sample B.

Referring to the example given earlier, we see that

$$U_a = 9 + \frac{12}{2} - 14 = 1$$

$$U_b = 9 + \frac{12}{2} - 7 = 8$$

Thus, the value of U is one, as before.

25.3 PROCEDURE

(1)   Two random samples are independently drawn from the population under study, and these are subjected to two different experimental treatments. Let n_a be the size of one sample, n_b the size of the other, and N the sum of n_a plus n_b.

(2)   At the conclusion of the experiment, a single criterion measure is administered to all subjects. The measures from the two samples are combined in a single ordered series and ranked from one to N, using the subscripts a and b to identify each rank as belonging to a particular sample.

(3)   Determine the values of U_a and U_b in the fashion described in the previous section. Let U equal the smaller of these.

(4)   The calculated U is compared to the tabled value at the desired level of significance and for the appropriate sample sizes. A one- or two-tailed test may be used. If the calculated value is equal to or *smaller* than the tabled value, the hypothesis is rejected; the observed differences in the ranks are significant. The two experimental treatments are determined to produce different results at the specified level of significance, and the treatment which produced the higher

ranks is usually regarded as superior, although the exact interpretation is dependent upon the nature of the criterion variable. If the calculated value is *larger* than the tabled value, the hypothesis is retained, and no significant difference in the ranks has been established.

Table 8 in the Appendix uses the symbol m to represent the size of the larger sample and the symbol n to represent the size of the smaller sample. If the samples are the same size, the investigator may use either of these for n_a and the other for n_b. For example, if m equals 30 and n equals 20, the critical value of U for a one-tailed test at the .05 level is 216. Any value of U equal to or smaller than 216 is significant for the one-tailed test at the .05 level with samples of these sizes.

25.4 EXAMPLE

The experiment which served to illustrate the use of the t-test for two independent samples (see Section 23.4) will also be used to illustrate the use of the Mann-Whitney U-test.

(1) H: No difference in the ranks for the two diets.
     A: Higher ranks are associated with one of the diets.

(2) Mann-Whitney U-test, $n_a = n_b = 10$.

(3) .05 level, two-tailed test.
     R: $U \leqslant 23$.

(4)

All scores	All ranks	A-ranks	B-ranks
1a	1a	1	
2a	2a	2	
3a	3a	3	
4a	5a	5	
4a	5a	5	
4b	5b		5
5a	7.5a	7.5	
5a	7.5a	7.5	
6b	9b		9
7b	10.5b		10.5
7b	10.5b		10.5

All scores	All ranks	A-ranks	B-ranks
8a	13a	13	
8b	13b		13
8b	13b		13
9a	16a	16	
9a	16a	16	
9b	16b		16
10b	18.5b		18.5
10b	18.5b		18.5
11b	20b		20

$$\Sigma R_a = 76 \qquad \Sigma R_b = 134$$

Notice that in the case of tie scores, the ranks are evenly divided between them. For example, the 18th and 19th ranks are evenly divided for the two scores of 10.

$$U_a = 100 + \frac{110}{2} - 134 = 21$$

$$U_b = 100 + \frac{110}{2} - 76 = 79$$

Checking the calculations:

$$U_a + U_b = n_a n_b$$
$$21 + 79 = 100$$
$$U = 21$$

(5)   Reject the null hypothesis. There is a significant difference in the ranks of the two samples. The data suggest that Diet B secures greater weight gains than Diet A.

25.5   DISCUSSION

The Mann-Whitney U-test is one of the more popular nonparametric tests. It is a suitable alternative to the t-test for two independent samples when the assumptions underlying the t-test cannot be met. The U-test is almost as powerful as the t-test (about 95 percent relative power with typical research samples) and does not require homogeneity of variance nor normality of distribution. It does, however, have the assumption of continuity of distribution, and when large numbers of tie scores are encountered, it is much less powerful. It is especially appropriate for use with small samples where there is the greatest danger from violating the assumptions underlying the equivalent t-test.

For samples larger than those provided for in Table 8, a normal approximation is available:

$$z = \frac{U - \dfrac{n_a n_b}{2}}{\sqrt{\dfrac{n_a n_b (n_a + n_b + 1)}{12}}}$$

This value of z may be referred to in Table 3 in the Appendix to determine its probability of occurrence, or it may be referred to in Table 20.1 in the text for hypothesis testing. In the latter case, the hypothesis is rejected if the calculated z equals or exceeds the tabled value.

EXERCISES

25.1   Refer to Exercise 23.1. Substitute the Mann-Whitney U-test for the t-test for two independent samples.

25.2   Refer to Exercise 23.2. Substitute the Mann-Whitney U-test for the t-test for two independent samples.

25.3   Given the following independent random samples which have been subjected to experimentation, and in which the individuals have been rated according to the scores recorded:

Sample I: 38, 39, 44, 47, 50, 51, 52, 59, 60, 61, 73, 74, 78, 84, 90
Sample II: 42, 43, 54, 62, 67, 69, 70, 75, 80, 81, 86, 89, 91, 97, 98

Use the Mann-Whitney U-test to determine whether there is a significant difference in the ranks of the two samples. Use a two-tailed test at the .05 level.

THE WILCOXON
MATCHED-PAIRS
SIGNED-RANKS TEST

26.1 FUNCTION

The Wilcoxon matched-pairs signed-ranks test is a nonparametric alternative to the t-test for two related samples. It may be used in either repeated measurements or matched-pairs types of designs. For a discussion of these designs, see Section 24.1. The Wilcoxon test requires data on at least an ordinal scale, and this data is assumed to be continuously distributed. The test does not require normality of distribution.

26.2 RATIONALE

Suppose that a random sample of paired measures is available from some population of interest. Let d_i be the difference between any pair of measures. Rank these difference scores from one to N (where N is the number of pairs), with respect to magnitude but without respect to sign (for example, 0, +1, −2, −3, +4, etc.) After ranking the difference scores in this fashion, separate the ranks into two groups, those corresponding to the positive difference scores, and those corresponding to the negative difference scores. Let T_a be the sum of the ranks for the positive differences, T_b the sum of the ranks for the negative differences, and let T be equal to the smaller of these two.

There are 2^N unique sets of signed ranks in the situation described. If the relationship between the scores in each pair is a completely random one, each of these 2^N sets is equally probable. If N is six, for example, there are 64 sets, and the probability that T_a will be zero is 1/64; the probability that it will be one is also 1/64, and the probability that it will be one or zero is 2/64. The probability that T_b will be zero is 1/64, and the probability that either T_a or T_b will be zero is 2/64. Following this pattern, the sampling distribution of T could be established for any sample size, and a table could be constructed for testing both one- and two-tailed hypotheses at any desired levels of significance. Table 9 in the Appendix is such a table.

If the relationship between the scores in each pair is a completely random one, the expected values of T_a and T_b would be the same, and the value of T would be maximum under these circumstances. If, however, there is a systematic tendency for the positive differences to be greater than or less than the negative differences, T will tend to be smaller, with T equal to zero representing two maximally different samples.

26.3 PROCEDURE

(1)   The procedure for conducting the experiment is largely determined by the experimental design (see Section 23.1). In any event, a collection of paired measures is produced by the experiment.

(2)   Let d_i be the difference between any pair of scores. Rank these differences from one to N with respect to their absolute magnitudes (without respect to sign). For example, consider this collection where N is eight:

Difference: -1 -2 $+4$ -6 $+7$ $+8$ $+10$ -12
Ranks: 1   2   3   4   5   6   7   8

(3)   Separate the ranks into two groups, those corresponding to the positive differences, and those corresponding to the negative differences. Take the sum of each of these two groups of ranks, and let T equal the smaller sum. In the example above, the sum of the ranks for the positive differences is 21; for the negative differences it is 15; T is 15.

(4)   The value of T is compared to the tabled values at the desired level of significance and for the appropriate value of N. A one- or two-tailed test may be used. If the calculated value is equal to or smaller than the tabled value, the hypothesis is rejected, and the observed differences in the ranks are significant. The exact interpretation of a significant finding is dependent upon the experimental design used and the nature of the dependent variable. If the calculated value is larger than the tabled value, the hypothesis is retained, and no significant difference in the ranks has been established.

26.4 EXAMPLE

The experiment that served to illustrate the use of the t-test for two related samples (see Section 24.4) will also be used to illustrate the Wilcoxon matched-pairs signed-ranks test.

(1) H: No difference in the ranks for the two methods.
     A: Higher ranks are associated with one of the methods.

(2) Wilcoxon matched-pairs signed-ranks test, $N = 10$.

(3) .05 level, two-tailed test.
     R: $T \leqslant 8$.

(4)

A-scores	B-scores	d_i	Rank	Positive	Negative
12	16	4	7.5	7.5	
11	9	-2	3.5		3.5
12	15	3	5.5	5.5	
10	10	0	0.0		
10	9	-1	1.5		1.5
9	12	3	5.5	5.5	
7	9	2	3.5	3.5	
6	10	4	7.5	7.5	
4	9	5	9.0	9.0	
4	3	-1	1.5		1.5

$$T_a = 38.5 \qquad T_b = 6.5$$

$T = 6.5$, but notice that one of the difference scores is zero and not ranked. Because it is neither positive nor negative, it is discarded from the analysis, and N is reduced to 9. A new region of rejection must be determined from Table 9. R: $T \leqslant 6$

(5) Retain. No significant difference in the ranks was found. This is consistent with the earlier finding using the t-test.

26.5 DISCUSSION

An important assumption underlying the Wilcoxon test and many other nonparametric statistics is that of continuity of distribution. The presence of tie scores suggests that this assumption has been violated. Whether or not this is of any consequence depends upon the manner of calculating the test statistic. There are two kinds of ties possible with the Wilcoxon, and both are illustrated in the example quoted: (1) tie scores for a given pair resulting in a zero difference, and (2) tie ranks resulting from pairs having the same differences. The first type of tie is of great importance. The usual recommendation is to discard these from the analysis and reduce N accordingly. This procedure cannot be justified, since the presence of these ties is evidence for retention of the null hypothesis. If there is an even number of these ties, they may be evenly divided among the positives and negatives; if there is an odd number, one may be discarded and the remainder divided. A better procedure might call for randomly assigning them to the positives and negatives. An even more acceptable procedure would be to assign them to the algebraic sign most conducive to retention of the null hypothesis. Each of these suggestions is based upon the assumption that the data are truly continuous but crudely measured. None of these is a very satisfying solution, and the presence of this type of tie is a strong argument against the use of this statistical test. The second type of tie is less of a problem, since the corresponding ranks may be evenly divided among the tied differences.

The Wilcoxon matched-pairs signed-ranks test has been suggested as a nonparametric alternative to the t-test for two related samples when the underlying assumptions for the latter test cannot be met. The relative power of the Wilcoxon is on the order of 95 percent with small samples and somewhat less with larger ones. The Wilcoxon, however, is not without assumptions of its own—the assumption of continuity of distribution giving rise to the problems noted in the previous paragraph.

The student's attention is directed to the fact that the assumptions underlying the t-test for related measures are not nearly as likely to be a problem as those underlying the t-test for independent samples. This fact, coupled with the problems encountered with tie scores in the Wilcoxon test, suggests that the Wilcoxon will not be nearly so popular as the Mann-Whitney in behavioral research.

EXERCISES

26.1   Refer to Exercise 24.1. Substitute the Wilcoxon matched-pairs signed-ranks test for the t-test for related samples.

26.2   Refer to Exercise 24.2. Substitute the Wilcoxon matched-pairs signed-ranks test for the t-test for related samples.

THE CHI-SQUARE DISTRIBUTION

27.1 THE CONTINUOUS DISTRIBUTION OF CHI-SQUARE

Consider the situation in which a very large number of random samples of the same size are drawn from a normally distributed population. If each measure is expressed as a relative deviation from the mean (a z-score), and if the sum of the squares of these relative deviations is obtained for each sample, the sampling distribution of these sums of squares, first derived by Karl Pearson, is known as the *chi-square distribution*. These ideas may be expressed symbolically by:

$$\chi^2 = \sum \left(\frac{X_i - \mu}{\sigma}\right)^2 = \Sigma z_i^2$$

Like the t-distribution, the chi-square distribution is a function of sample size; there is a family of chi-square curves, each determined by a single parameter, its degrees of freedom. The term "degrees of freedom" refers to the number of independent squares in the sum. Unlike the t-distribution, the chi-square distribution is sharply skewed to the right for small samples; however, as N increases the distribution tends toward symmetry and becomes nearly bell-shaped for large samples. All chi-square values are positive, and they range continuously from zero to infinity. The mean of any chi-square distribution is equal to its degrees of freedom, and the mode of each distribution is at $df - 2$ except for the curve with one degree of freedom. The distribution of chi-square for representative values of degrees of freedom is graphically illustrated in Fig. 27.1.

Chi-square distributions with 30 or more degrees of freedom tend to have essentially the same form, and it is customary to tabulate the

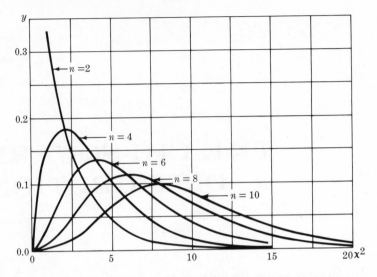

Figure 27.1   Representative Chi-Square Distributions.

statistic for degrees of freedom from one to 30. Because chi-square distributions are skewed, and the mean is different for each distribution, the method of presenting the distributions in tabular form differs from that used with the t-distributions. The student's attention is directed to Table 10 in the Appendix. The probabilities recorded across the top of the table are those associated with random chi-squares equal to or larger than the tabled values. Most applications of the chi-square statistic will call for a one-tailed test, and only those values recorded on the right-hand side of Table 10 will be used. Any probability recorded at the top of the table may be chosen for the level of significance. In the event that a two-tailed test is desired, the determination of the region of rejection requires a little more effort. Suppose a two-tailed test at the .01 level is desired, and the number of degrees of freedom is 30. The region of rejection will include all chi-squares equal to or smaller than 13.8 ($P = .995$) plus all chi-squares equal to or greater than 53.7 ($P = .005$).

27.2 NORMAL APPROXIMATION OF THE CHI-SQUARE DISTRIBUTION

The behavioral scientist is unlikely to encounter many situations in which he will want to test the significance of sample chi-squares with more than 30 degrees of freedom; however, if the need arises, a normal approximation may be used. It has been established that, as the value of N

increases, the distribution of $\sqrt{2\chi^2}$ approaches the normal distribution with mean of $\sqrt{2df - 1}$ and standard deviation of unity. One may obtain a good approximation of the probability of any chi-square with 30 or more degrees of freedom by calculating the normal deviate:

$$z = \sqrt{2\chi^2} - \sqrt{2df - 1}$$

For example, Table 10 reports a chi-square of 43.77 for 30 degrees of freedom at the .05 level. Substituting this information in the formula given above, we obtain:

$$z = \sqrt{87.44} - \sqrt{59} = 9.35 - 7.68 = 1.67$$

This value may be referred to in Table 3, where its probability is determined to be approximately .05.

27.3 TEST OF A HYPOTHESIS ABOUT THE VARIANCE

Consider the situation in which a random sample is drawn from a normally distributed population, and the investigator wishes to test the hypothesis that the population variance is equal to some specified value, σ^2. The unbiased estimate of the parameter may be obtained from:

$$S^2 = \frac{SS}{N - 1}$$

Notice that $SS = (N - 1)S^2$.

We have previously defined the chi-square statistic as:

$$\chi^2 = \left(\frac{X - \mu}{\sigma}\right)^2 = \frac{SS}{\sigma^2}$$

The degrees of freedom for this chi-square will be $N - 1$ if the sum of squares is calculated using the sample mean as an estimate of the population, which will be the case.

Then, with a little simple algebra, one may demonstrate that:

$$\chi^2 = \frac{(N - 1)S^2}{\sigma^2}$$

This value may be compared to the tabled value with the appropriate number of degrees of freedom and the desired level of significance to determine whether the sample value departs significantly from the hypothesized value. This is probably the only situation in which the reader will encounter a two-tailed test with the chi-square statistic.

Suppose, for example, that an investigator wishes to determine whether the IQs of the students in a certain school district are more or less variable than those of the norm group ($\sigma^2 = 225$). If a two-tailed test at the .05 level is desired a chi-square equal to or smaller than 16.8 or a chi-square equal to or greater than 47.0 will be significant. Let us suppose that the variance estimate from the sample data is 360:

$$\chi^2 = \frac{(30 - 1)360}{225} = 46.4$$

The verdict is to retain the null hypothesis; the available evidence does not establish that the variance of the population under study differs significantly from that of the norm group.

27.4 CHI-SQUARE APPROXIMATION OF THE MULTINOMIAL DISTRIBUTION

The general equation of the binomial probability distribution has been previously identified as:

$$P(r \text{ in } n) = \binom{n}{r} p^r q^{n-r}$$

Suppose we effect a change in notation such that n_1 is substituted for r, n_2 for $n - r$, N for n, P_1 for p, and P_2 for q. Thus, the equation becomes:

$$P(n_1 \text{ in } N) = \binom{N}{n_1} (P_1)^{n_1}(P_2)^{n_2} \quad \text{where} \quad \binom{N}{n_1} = \frac{N!}{n_1! n_2!}$$

This notation is convenient for extending the expression from binomial populations to multinomial populations. Consider, for example, the expression extended to a trinomial population:

$$\frac{N!}{n_1! n_2! n_3!} (P_1)^{n_1}(P_2)^{n_2}(P_3)^{n_3}$$

This expression may be extended to as many categories as are needed. The student should readily see that the ability to test hypotheses about nominal variables with no restriction on the number of categories (such as we encountered with the binomial) would be a very valuable asset indeed. Unfortunately, the calculation of multinomial probabilities, even with very small samples, is laborious and often prohibitive. Furthermore, these calculations yield point probabilities, and the determination of the region of rejection for testing a given hypothesis is likely to require many such calculations.

Consider, for example, the very simple situation in which three categories are used, the three probabilities are equal ($P_1 = P_2 = P_3 = 1/3$), and the sample size is 3. The theoretical sampling distribution of the three objects is given in Table 27.1.

Table 27.1   Multinomial Distribution with $N = 3$

Distribution						Probability
0	0	3	1	$(1/3)^0$	$(1/3)^0$	$(1/3)^3 = 1/27$
0	1	2	3	$(1/3)^0$	$(1/3)^1$	$(1/3)^2 = 3/27$
1	0	2	3	$(1/3)^1$	$(1/3)^0$	$(1/3)^2 = 3/27$
0	2	1	3	$(1/3)^0$	$(1/3)^2$	$(1/3)^1 = 3/27$
1	1	1	6	$(1/3)^1$	$(1/3)^1$	$(1/3)^1 = 6/27$
0	3	0	1	$(1/3)^0$	$(1/3)^3$	$(1/3)^0 = 1/27$
1	2	0	3	$(1/3)^1$	$(1/3)^2$	$(1/3)^0 = 3/27$
2	0	1	3	$(1/3)^2$	$(1/3)^0$	$(1/3)^1 = 3/27$
2	1	0	3	$(1/3)^2$	$(1/3)^1$	$(1/3)^0 = 3/27$
3	0	0	1	$(1/3)^3$	$(1/3)^0$	$(1/3)^0 = 1/27$

$$27/27$$

As the number of categories and the sample size are increased, not only do the calculations expand unendingly, but the identification of a useful region of rejection becomes increasingly impractical. If the sample size is adequate, both of these difficulties may be overcome through the use of the chi-square approximation of the multinomial, which is equivalent to the normal approximation of the binomial. The next two chapters of this text are devoted to this application of the chi-square statistic.

CHI-SQUARE TESTS OF
GOODNESS OF FIT

28.1 FUNCTION

The behavioral scientist often encounters research problems in which the variables are expressed as the number of individuals, objects, or events falling in each of several categories. We have been referring to these as *nominal* variables. When there are only two categories, the binomial probability distribution provides a useful model for testing some hypotheses involving nominal variables; however, when there are more than two categories, the binomial model is no longer applicable. The chi-square statistic provides a generalized procedure for testing hypotheses about the distributions of nominal and higher order data.

Chi-square tests of goodness of fit, sometimes called *one-sample chi-square tests*, are used to determine whether an observed frequency distribution departs significantly from a hypothesized frequency distribution. These tests perform essentially the same function as the binomial probability distribution in hypothesis testing, but the chi-square tests of goodness of fit are not restricted to two categories.

28.2 RATIONALE

The binomial probability distribution is a special case of the multinomial probability distribution, the binomial being limited to two categories, the multinomial having no such limitation. While the multinomial probability distribution is a useful model for certain statistical analyses, its calculation tends to be laborious and often prohibitive, and it yields point probabilities which are not readily combined in a fashion useful to the behavioral scientist. For these reasons, the multinomial probability distribution is seldom used by the behavioral scientist. The chi-square statistic, however, provides a relatively simple and extremely useful approximation of the multinomial which is equivalent to the normal approximation of the binomial.

Typically, the number of degrees of freedom in a chi-square goodness of fit test will be equal to the number of categories (cells) less one. The loss of one degree of freedom is due to the restriction that the sum of the expected frequencies must equal the sum of the observed frequencies. This is equivalent to specifying the frequency in the last cell; it is not free to vary.

28.3 PROCEDURE

(1) The application of a chi-square goodness-of-fit test greatly resembles the use of the binomial probability distribution in hypothesis testing. A theoretical distribution is either hypothesized or is known to exist in the completely random situation. That is, the investigator must establish a hypothetical frequency distribution based upon whatever information he may have. Three examples will be considered: (a) the investigator might hypothesize that the population is uniformly distributed on the trait under study, (b) he might hypothesize that the population is normally distributed on the trait under study, or (c) he might hypothesize a distribution structured around scientific speculation, such as a hypothesis about the distribution of a hereditary trait. The categories may be defined on a nominal, ordinal, or interval scale, but the categories must be mutually exclusive and collectively exhaustive.

(2) The common practice is to represent each category as a cell in a table such as this one containing four cells:

1   2   3   4

The theoretical (or expected) frequencies will be placed in the small boxes in the upper right corners of the cells; the observed frequencies will be placed in the main parts of the cells.

(3) Expected frequencies are calculated as follows: Let p_j be the probability (theoretical relative frequency) of an object falling in the jth

cell. Then the expected frequency for the jth cell (E_j) is equal to $p_j N$, where N is the sample size. Of course, the sum of the p_j's must equal one, and the sum of the expected frequencies must equal N.

(4) A random sample of objects is drawn from the population being studied, and the frequency distribution of the sample is determined. The observed frequencies are recorded in their respective cells.

(5) The chi-square is calculated as follows:

$$\chi^2 = \sum_{j=1}^{k} \frac{(O_j - E_j)^2}{E_j}$$

where O_j = the observed frequency for the jth cell
E_j = the expected frequency for the jth cell
k = the number of cells
$df = k - 1$

(6) If the number of degrees of freedom is one, a continuity correction will be used in the calculation of chi-square:

$$\chi^2 = \sum_{j=1}^{k} \frac{(|O_j - E_j| - \frac{1}{2})^2}{E_j}$$

(7) At some predetermined level of significance, the calculated chi-square is compared with the tabled value having $k - 1$ degrees of freedom. Only a one-tailed test is appropriate. If the calculated statistic equals or exceeds the tabled value, the finding is significant, and the hypothesis of no difference between the population distribution and the hypothesized distribution is rejected: the population distribution is determined to differ from the hypothesized distribution at the specified level of significance. If the calculated value is smaller than the tabled value, the hypothesis is retained, and no significant difference in the two distributions has been determined to exist.

28.4 EXAMPLE

A high school algebra teacher suspects that his students tend to be either especially bright or especially stupid. A standardized algebra test is administered to 60 students, and their scores are recorded as percentile ranks based on national norms. The percentile ranks were, of course, uniformly distributed in the norm group. The teacher hypothesizes that his students are not distributed in the same fashion as the norm group. The test of the hypothesis is outlined below:

(1) H: The students are uniformly distributed.
     A: The students are not uniformly distributed.

(2) Chi-square goodness of fit, $N = 60$.

(3) .05 level, $df = 9$ (10 cells will be used).
     R: $\chi^2 \geqslant 16.9$

(4) The expected frequencies in a uniform distribution are all the same and may be calculated from: $E_j = N/k$. With a sample of size 60 evenly divided among 10 cells, each expected frequency will be 6. The 10 cells will be partitioned at the 10th, 20th, 30th, 40th, 50th, 60th, 70th, 80th, and 90th centiles. A frequency distribution is constructed with the data grouped into 10 classes with the class limits just described. Both the expected and observed frequencies are recorded in the table below:

6	6	6	6	6	6	6	6	6	6
9	10	4	2	3	5	1	7	8	11

| | 10 | 20 | 30 | 40 | 50 | 60 | 70 | 80 | 90 | |

The chi-square is calculated from:

$$\chi^2 = \sum \frac{(O_i - E_i)^2}{E_i} = \frac{9}{6} + \frac{16}{6} + \frac{4}{6} + \frac{16}{6} + \frac{9}{6} + \frac{1}{6}$$
$$+ \frac{25}{6} + \frac{1}{6} + \frac{4}{6} + \frac{25}{6} = \frac{110}{6} = 18.3$$

(5) Reject the null hypothesis. The sample deviates significantly from the hypothesized uniform distribution. The data appear to bear out the teacher's suspicions; however, rejection of the null hypothesis does not necessarily establish that the students are concentrated in the extreme cells; it establishes only that they are not distributed in the same fashion as the norm group.

28.5 CHI-SQUARE TEST FOR GOODNESS OF FIT TO NORMAL

The procedure used to test for uniformity of distribution requires only slight modification to test for normality of distribution. In this situation, the normal deviates (z-scores) corresponding to the percentile ranks are determined from Table 3 in the Appendix, and these are used for the class limits. If the sample mean and standard deviation are used for the calculation of standard scores for the sample, two additional degrees of freedom are lost ($df = k - 3$). If the instrument is scaled to yield standard scores based on national norms, and these are used to determine the class limits, the degrees of freedom are as before ($df = k - 1$), but we are now testing a hypothesis of normality with specified mean and standard deviation. Conceivably, the population under study might be normally distributed, and yet the hypothesis might be rejected due to difference in means or standard deviations. The use of chi-square goodness-of-fit procedures to test for normality is not recommended for samples smaller than 50 in size.

28.6   DISCUSSION

Chi-square goodness-of-fit tests are useful procedures for testing whether a distribution of measures deviates significantly from a hypothesized distribution.   The data may be recorded on nominal or higher order scales.

The use of the chi-square statistic in goodness-of-fit tests is an approximation of the multinomial probability distribution and carries with it an important restriction with respect to sample size. In order for this continuous model to provide an adequate approximation of a discrete distribution, fairly large samples are required. Research has determined that a good approximation is achieved when at least 80 percent of the cells have expected frequencies equal to five or larger.

The requirements with respect to minimum expected frequencies have proved to be a severe handicap in the use of the chi-square statistic. However, a number of alternatives are available. Often, the investigator will be able to justify the combining of classes to increase the expected frequencies. Of course, there must be some logical basis for combining two cells—they must have something in common. The binomial probability distribution has no restrictions with respect to minimum expected frequencies and may be substituted for the chi-square in some goodness-of-fit tests. If the data are ordinal and continuously distributed, the Kolmogorov statistic may be the appropriate alternative. However, its use is not recommended if the data are grouped or if there are many tie scores.

EXERCISES

28.1   An ESP experiment is conducted in which the probability of a right answer by chance is $1/4$. In 36 trials, a certain subject got 16 right. Test the hypothesis that he got this many answers right by chance. Use a chi-square goodness-of-fit test at the .01 level with the correction for continuity. Then calculate the probability of getting this many right using the normal approximation of the binomial.

28.2   Given the following collection of scores which were randomly drawn from a much larger collection:

81	73	67	61	54	51	44	39	33	21
80	73	65	60	54	50	43	37	32	20
78	72	64	58	54	48	43	36	32	20
76	71	64	57	54	48	43	34	31	20
75	69	62	57	53	47	41	33	28	19
74	69	61	55	52	47	40	33	25	19
73	68	61	55	52	46	40	33	23	18

Test the hypothesis that the parent population is normally distributed with mean of 50 and standard deviation of 10. Use the .05 level of significance.

28.3   A high school genetics student hypothesizes that the offspring of a certain pair of guinea pigs will be distributed in the ratio 9:3:3:1 with respect to color and texture of coat. He predicts that 6.25 percent of the offspring will have smooth white coats, 18.75 percent smooth black coats, 18.75 percent rough white coats, and 56.25 percent rough black coats. A total of 96 progeny are raised from this pair, and they include 4 with smooth white coats, 14 smooth black, 23 rough white, and 55 rough black. Test this hypothesis at the .05 level.

CHI-SQUARE TESTS OF INDEPENDENCE

29.1 FUNCTION

Chi-square tests of independence are extremely useful statistical procedures for determining whether two nominal (or higher level) measures are related. If one of the variables is group membership and the other a criterion of some sort, the test may be used to determine whether two or more populations are distributed in the same fashion with respect to the criterion.

The data are organized into a bivariate frequency table, or contingency table as it is sometimes called, and the statistical test is made to determine whether classification on the row variable is independent of classification on the column variable. For example, suppose the row variable were used to classify the subjects with respect to political affiliation and the column variable with respect to religious affiliation. The chi-square test of independence is used to determine whether there is a relationship between the two nominal variables. We seek to answer the question "With respect to political affiliation, do Protestants and Catholics represent two different populations or a single population, or, with respect to religious affiliation, do Republicans and Democrats represent two different populations, or a single population?"

There is no restriction with respect to the number of categories when the chi-square statistic is used in tests of independence; however, there is the usual limitation with respect to sample size. Generally, the use of the chi-square statistic requires rather large samples, especially if the number of categories is large.

29.2 RATIONALE

Chi-square tests of independence are an extension of the chi-square approximation of the multinomial to the situation in which a double classification system is used and the expected frequencies are derived from sample data.

Let us suppose, for example, that each of the persons in a certain population can be classified as either Republican or Democrat, and that each can also be classified as either Catholic or Protestant. A random sample of 100 persons is drawn, and it is determined to be evenly divided with respect to political affiliation, there being 50 Democrats and 50 Republicans. The sample is also determined to have 40 Catholics and 60 Protestants. If the two classification systems are independent of each other, we would expect the 40 Catholics to be evenly divided with respect to political affiliation, and we would also expect the 60 Protestants to be evenly divided with respect to political affiliation. The expected frequencies are 20 Catholic Republicans, 20 Catholic Democrats, 30 Protestant Republicans, and 30 Protestant Democrats. The situation is illustrated in the table below:

	Republicans	Democrats	
Catholics	20	20	40
Protestants	30	30	60
	50	50	100

Notice that no sample information has been recorded other than the marginal frequencies (the marginal frequencies are the row and columns sums). The expected frequencies are derived from the marginal frequencies. These expected frequencies may be calculated from the formula:

$$E_{ij} = \frac{R_i C_j}{N}$$

where E_{ij} = the expected frequency for cell in row i, column j
R_i = the sum of the frequencies in row i
C_j = the sum of the frequencies in column j
N = the sum of the frequencies for all cells

A quick check on the calculations will verify that the sums of the expected frequencies in each row and column are equal to the marginal frequencies. These restrictions determine the number of degrees of freedom to be

$(r - 1)(c - 1)$, where r is the number of rows and c is the number of columns. With a little effort, the student should be able to see that there is one cell in each row and column which is not free to vary. In the example above, the number of degrees of freedom is one.

The chi-square statistic is calculated by summing over all cells:

$$\chi^2 = \sum \sum \frac{(O_{ij} - E_{ij})^2}{E_{ij}}$$

29.3 PROCEDURE

(1)   Two methods of sampling are commonly used in chi-square tests of independence: (a) A single random sample may be drawn from a population that can be classified on two different variables. The two variables become the row and column variables in a bivariate frequency table. (b) A random sample may be drawn from each of two or more populations, all of which can be classified on the same variable. In this situation, the various populations will be used as the second variable in constructing the bivariate frequency table. The analysis is the same in either case, and often the interpretation is the same.

(2)   A bivariate frequency table is constructed, and the marginal frequencies are determined. The expected frequencies are calculated in the fashion described in the previous section using the formula:

$$E_{ij} = \frac{R_i C_j}{N}$$

(3)   The observed frequencies are recorded in the table, and the chi-square statistic is calculated from the formula:

$$\chi^2 = \sum \sum \frac{(O_{ij} - E_{ij})^2}{E_{ij}}$$

(4)   In the case of a two-by-two table, where the number of degrees of freedom is one, a continuity correction should be used. In this case, the formula for the chi-square would be:

$$\chi^2 = \sum \sum \frac{(|O_{ij} - E_{ij}| - \frac{1}{2})^2}{E_{ij}}$$

Some computational effort may be saved if the data are organized into a two-by-two table in which A is the frequency in cell$_{11}$, B the frequency in $cell_{12}$, C the frequency in $cell_{21}$, and D the frequency in $cell_{22}$. The formula becomes:

$$\chi^2 = \frac{N(|AD - BC| - N/2)^2}{(A + B)(C + D)(A + C)(B + D)}$$

(5)   At some predetermined level of significance, the calculated chi-square is compared with the tabled value having $(r - 1)(c - 1)$ degrees of freedom. Only a one-tailed test is appropriate. If the calculated statistic equals or exceeds the tabled value, the finding is significant and the hypothesis of independence is rejected; the row and column variables are determined to be dependent at the specified level of significance. The exact interpretation will be dependent upon the nature of the row and column variables. If the calculated statistic is smaller than the tabled value, no significant relationship between the row and column variables has been determined to exist.

29.4   EXAMPLE

A group of college students is asked to answer the following multiple-choice question in a values inventory:

What do you believe is the true nature of God?

(1)   I believe in a personal God who has revealed Himself in the Bible.
(2)   There is a God, Father of all men, who is common to all religious faiths. It is not particularly important whether a man is a Christian, Jew, Moslem, Hindu, etc.
(3)   I believe in a Supreme Being or First Cause, but I cannot believe in a personal God.
(4)   The nature of God is not (or cannot) be known by man.
(5)   There is no God.

The respondents were also identified by sex, and the following bivariate frequency table was constructed from the data:

Responses:	1		2		3		4		5		
Males	18		40		7		4		1		70
	12		33		17		8		0		
Females	36		80		14		8		2		140
	42		87		4		4		3		
	54		120		21		12		3		210

Notice that when the expected frequencies were calculated, three of the ten cells had expected frequencies less than five. In order to meet the criterion of 80 percent of the cells having expected frequencies of five or more, it is necessary to collapse the adjacent cells where the frequencies are low. An examination of the inventory question and its responses suggests that this is a logical thing to do. Thus a new bivariate frequency table is constructed in which responses 4 and 5 are combined into a single category:

Responses:	1	2	3	4-5	
Males	18 / 12	40 / 33	7 / 17	5 / 8	70
Females	36 / 42	80 / 87	14 / 4	10 / 7	140
	54	120	21	12	210

The test of the hypothesis is outlined below. The investigator would probably choose to think of sex as the independent variable and response as the dependent variable. He is seeking to determine whether the sex of the respondent influences the response to the question.

(1) H: Sex of respondent and response are independent.
    A: Response is dependent upon the sex of the respondent.

(2) Chi-square test of independence, $N = 210$, 2×4 table.

(3) .01 level, $df = 3$.
    R: $\chi^2 \geqslant 11.3$.

(4) The calculation of the expected frequencies will be illustrated with that for $cell_{11}$:

$$E_{11} = \frac{R_1 C_1}{N} = \frac{(70)(54)}{210} = 18$$

The calculation of the chi-square is as follows:

$$\chi^2 = \frac{36}{18} + \frac{49}{40} + \frac{100}{7} + \frac{9}{5} + \frac{36}{36} + \frac{49}{80} + \frac{100}{14} + \frac{9}{10}$$
$$= 27.5$$

(5) Reject the null hypothesis. There is a significant relationship between sex of the respondent and response to the question. While the test of independence does not specifically identify the difference between the responses of the two groups, an examination of the

available responses and of the bivariate frequency table suggests that the females are more inclined to believe in a personal God than are the males.

29.5 THE MEDIAN TEST

The median test has been suggested as a nonparametric alternative to the t-test for two independent samples when the assumptions underlying both the t-test and the Mann-Whitney U-test cannot be met. It is simply an application of the chi-square test of independence to the situation in which the investigator has ordinal data and wishes to determine whether there is a significant difference in the scores of two or more samples. The scores for all samples are arranged in a single ordered series and their common median determined (see Section 7.4 for instructions for calculating the value of the median). Then a bivariate frequency table is constructed with one column for each sample and with two rows; the top row is for individuals who scored above the median, the bottom for individuals who scored below the median. The number of persons in each sample who scored above and below the common median are entered in the appropriate cells in the table, and the expected frequencies and chi-square statistic are calculated as in the previous section. Consider the following table:

Group:	I	II	III	
Above	8 / 12	10 / 6	9 / 9	27
Below	8 / 4	10 / 14	9 / 9	27
	16	20	18	54

The chi-square for this data is 7.2 with two degrees of freedom, which is significant at the .05 level. Although one may safely conclude that Group I placed significantly more individuals above the common median than did Group II, the test of the hypothesis establishes nothing with respect to the relationship between Group I and Group III, nor with respect to the relationship between Group II and Group III. This is a problem common to the testing of hypotheses about more than two samples. Consider also the possibility that the individuals in Group I who scored above the median were just barely above the median, and that those in Group II who scored above the median were the highest scorers of all. Similarly, those in Group I who scored below the median could have been the lowest scorers of all, while those in Group II could have been just below the median. The possibility exists that the mean for Group II is actually higher than the mean for Group I, even though Group I placed more subjects above the common median. The investigator who uses the median test must carefully consider the hypothesis which he wishes to test and whether or not the median test actually tests that hypothesis.

Generally, the median test will not prove nearly as satisfactory as the Mann-Whitney U-test; however, it may be preferred when the measurement is crude, and it has the additional advantage of not being restricted to the two-sample situation.

29.6 THE CONTINGENCY COEFFICIENT

The contingency coefficient is an index of relationship between two variables for the situation in which the data are organized into a bivariate frequency table. It is the only correlation coefficient appropriate for use with nominal data, with the possible exception of the phi coefficient, which is limited to two-by-two tables.

The magnitude of a chi-square statistic calculated from a bivariate frequency table is a function of the relationship between the row and column variables. This fact is utilized in the calculation of the contingency coefficient from the formula:

$$C = \sqrt{\frac{\chi^2}{N + \chi^2}}$$

Whether the relationship between the two variables is a significant one may be tested in the same fashion as in the test of independence, referring the chi-square to Table 10.

Unfortunately, the size of the contingency coefficient is a function of the number of cells in the table, and under no circumstances is it possible for the coefficient to be unity, even though a perfect relationship may exist. The maximum value of the contingency coefficient for a two-by-two table is .707, for a three-by-three table is .816, for a four-by-four table .866, and for a five-by-five table .894. The maximum value for a ten-by-ten table is .949. In addition to the fact that it is not directly comparable with other correlation coefficients, the contingency coefficient requires the large sample sizes characteristic of the chi-square statistic, and for these reasons it has not proved very popular.

The author suggests a slight modification of the formula for the contingency coefficient so that a perfect relationship will produce a coefficient of unity:

$$C_r = C \sqrt{\frac{k}{k-1}} = \sqrt{\left(\frac{k}{k-1}\right)\left(\frac{\chi^2}{N+\chi^2}\right)}$$

The constant k is the number of cells in each column or row (a symmetrical frequency table is assumed). Thus, k equals four for a four-by-four table.

Cramér has also suggested an alternative to the contingency coefficient to produce a coefficient of unity for a perfect relationship. This is known as Cramér's statistic:

$$C_c = \sqrt{\frac{\chi^2}{N(L-1)}}$$

The constant L is the number of cells in each column or row, whichever is smaller. Thus, L equals three for a three-by-five table.

It should be noted that these two procedures do not yield the same coefficient if the relationship is imperfect. Both are regarded as superior to the traditional contingency coefficient; however, neither has had wide acceptance up to this time.

29.7 DISCUSSION

Chi-square tests of independence are by far the most valuable of the nonparametric procedures available to the behavioral scientist. They are used to test the significance of relationship between row and column variables in bivariate frequency tables. The data may be recorded on nominal or higher order scales.

The use of the chi-square statistic in tests of independence is an approximation of the multinomial probability distribution and carries with it the restriction with respect to sample size which is characteristic of continuous models used with discrete data. A satisfactory approximation is assured if at least 80 percent of the cells in the table have expected frequencies equal to five or more.

The requirements with respect to minimum expected frequencies have proved to be a severe handicap in the use of the chi-square statistic, especially in tests of independence. Often, it will not be practical to collapse adjacent cells, and some alternative statistical procedure is needed. If the data are ordinal and continuously distributed, the Kolmogorov-Smirnov statistic may be the appropriate alternative in the two-sample situation. Its use is limited to two samples, however, and it is not recommended for use with grouped data or in cases where there are many tie scores. In the case of two-by-two contingency tables, the Fisher exact test may be substituted for use with small samples, since it has no restrictions with respect to minimum sample size.

EXERCISES

29.1 A multiple-choice type of question with respect to the desirability of teacher tenure is given to several groups of interested persons. Three responses to the question were available: (a) agree, (b) no opinion, and (c) disagree. A group of teachers split on the question, with 75 choosing agree, 10 no opinion, and 5 disagree. A group of school administrators was divided on the issue, with 20 choosing agree, 10 choosing disagree, and none choosing no opinion. A group of businessmen was evenly divided on the issue, with 10 choosing each response. Construct a bivariate frequency table from the data, and test the hypothesis of independence at the .05 level.

29.2 Random samples of students are chosen from the public high school and the parochial high school of a certain community. These are then classified into five socio-economic classes according to parent's occupation. The 30 students from the parochial school included 2 whose fathers were classified professional or managerial, 0 semiprofessional, 12 skilled workers, 14 semi-skilled, and 2 unskilled. The 60 students from the public school were classified 4 professional or managerial, 9 semiprofessional, 18 skilled workers, 22 semiskilled, and 7 unskilled. Construct a bivariate frequency table from the data, and test the hypothesis of independence at the .05 level.

29.3 Refer to Exercise 25.3. Use the median test at the .05 level to determine whether there is a significant difference in the two groups.

29.4 A certain projective instrument is intended for use in classifying personality disorders, and scores yielded by the instrument are used to rate the subject as having Type A, B, C, or D personality. Two psychologists rate each of a group of 100 individuals who are undergoing therapy. The rating of one psychologist is used as the row variable and that of the other psychologist as the column variable in the following contingency table. Calculate the contingency coefficient for this data, and test the significance of relationship.

	A	B	C	D
A'	12	4	8	0
B'	7	18	2	3
C'	2	2	15	6
D'	0	1	3	17

TESTS OF SIGNIFICANCE
OF CORRELATION

30.1 INTRODUCTION

Occasionally, one reads or hears that a correlation coefficient on the order of .80 or .90 signifies a high degree of relationship, while a coefficient of .20 or .30 signifies a low relationship. In statistical inference, one cannot attach meaning to a correlation coefficient without taking into consideration the size of the sample and the sampling distribution of the coefficient. Consider, for example, the ten pairs of scores reported in Figure 12.5:

X-score: 0 2 3 4 4 5 5 8 9 10
Y-score: 3 8 5 6 5 4 4 7 1 7

A Pearson product moment correlation coefficient calculated from these ten pairs of scores is equal to zero. Suppose, however, that a sample of three pairs of scores was drawn at random from this collection, and that the three pairs happened to be:

X-score: 0 4 8
Y-score: 3 5 7

These three scores happen to lie on a straight line, and a Pearson coefficient calculated from them is equal to plus one. This suggests that a test of significance of the relationship between the variables is needed whenever the investigator draws conclusions about population correlations from sample data. As a general rule, the investigator will test the null hypothesis that the population correlation coefficient is zero.

30.2 THE PEARSON AND POINT BISERIAL CORRELATION COEFFICIENTS

The form of the sampling distribution of the Pearson correlation coefficient is a function of the magnitude and the sign of the coefficient in the population under study. As the magnitude of the coefficient increases, the sampling distribution becomes increasingly skewed—to the left for positive correlations and to the right for negative correlations. However, when the population value is zero, the sampling distribution is symmetrical and approximately normal, with the approximation improving with increased sample size. For this reason, the t-distribution is an appropriate model for testing the null hypothesis that the population correlation coefficient is zero, but not for testing the hypothesis of no difference between two correlation coefficients. The t-statistic may be calculated from:

$$t = r \sqrt{\frac{N - 2}{1 - r^2}} \quad \text{with } df = N - 2$$

The calculated value may be compared to the tabled t with $N - 2$ degrees of freedom, where N is the number of pairs of measures. A significant relationship exists if the observed value equals or exceeds the tabled value at the desired level of significance. A one- or two-tailed test may be used. The one-tailed test is often appropriate in learning research, since the investigator will generally be prepared to specify in advance that only a positive coefficient is of interest. However, if the investigator is interested in either a positive or negative coefficient, the two-tailed test must be used.

The formula given above may also be used to transform critical values of t to critical values of r, so that the region of rejection may be stated in terms of r instead of t. The formula for calculating critical values of r is:

$$r = \sqrt{\frac{t^2}{N - 2 + t^2}} \quad \text{with } df = N - 2$$

This formula has been used to construct Table 11 in the Appendix. A significant r is one equal to or larger than the tabled value with $N - 2$ degrees of freedom, where N is the number of pairs of scores. The table is complete enough for most behavioral research, but additional values may be calculated using the formula if they are needed. The significance of the point biserial coefficient is tested in precisely the same fashion as the Pearson.

The preceding paragraphs suggest the interesting possibility of substituting the point biserial or the Pearson coefficient for the t-statistic in experimental analysis. For example, a binary variable may be substituted for group membership in the t-test for two independent samples. Let us assign a zero to each member of the control group and a one to each member of the experimental group. These group scores are then paired with the criterion scores (giving n_1 plus n_2, or N pairs of scores), and a point biserial or Pearson correlation coefficient is calculated from the paired measures. When this correlation coefficient is tested for statistical significance, the result is precisely the same as the t-test for two independent samples. In addition to the test of significance, the investigator has an index of relationship between the two variables, and may meaningfully speak of the proportion of the variance of the criterion accounted for by membership in the two groups.

30.3   THE SPEARMAN RANK CORRELATION COEFFICIENT

When the Spearman rank correlation coefficient is calculated, each of the variables ranges from one to N. Suppose we arrange the X-ranks in order from one to N, and consider all possible pairings of the Y-ranks with these. There are $N!$ possible arrangements of the paired variables, and each of these is equally probable if the relationship between the two variables is a completely random one. For example, if N equals two, there are two possible combinations and two possible correlation coefficients, plus one and minus one. Each of these is equally probable, the probability being one-half. If we continued in this fashion, it would be a relatively simple though laborious process to develop the sampling distribution of the Spearman rank correlation coefficient for any sample size. Fortunately, this task has already been completed, and the results are recorded in Table 12 in the Appendix. A Spearman coefficient which equals or exceeds the tabled value is a significant one. For larger samples, the t-test used with the Pearson correlation coefficient may be used. A usable approximation is achieved with samples as small as 10 in size, and the approximation is excellent for samples as large as 30.

30.4   THE PHI COEFFICIENT AND THE
CONTINGENCY COEFFICIENT

The significance of either the phi coefficient or the contingency coefficient may be tested with a chi-square test of independence. In the case of the contingency coefficient, the calculation of the chi-square is an intermediate step in the determination of the correlation coefficient. The test of significance is discussed in Section 29.6.

In the case of the phi coefficient, the chi-square will ordinarily be calculated from the two-by-two frequency table with the correction for continuity (see Section 29.3 for the formula). If the sample size is quite large, the continuity correction may be omitted and the chi-square calculated from:

$$\chi^2 = N\phi^2$$

30.5 THE BISERIAL AND TETRACHORIC CORRELATION COEFFICIENTS

The sampling distribution of the biserial correlation coefficient is a function of the proportion of the subjects falling in each part of the dichotomy. This greatly complicates the problem of establishing a sampling distribution for the coefficient, and no completely satisfactory test of its significance is available. This is doubly true for the tetrachoric correlation coefficient.

EXERCISES

30.1   Refer to Exercise 12.1. Test the significance of the correlation coefficient. Use a one-tailed test at the .01 level.

30.2   Refer to Exercise 12.2. Test the significance of the correlation coefficient. Use a two-tailed test at the .05 level.

30.3   Refer to Exercise 12.3. Test the significance of the correlation coefficient. Use a two-tailed test at the .05 level.

30.4   Refer to Exercise 13.1. Test the significance of the correlation coefficient. Use a one-tailed test at the .05 level.

30.5   Refer to Exercise 13.5. Test the significance of the correlation coefficient. Use the .05 level.

KOLMOGOROV TESTS OF
GOODNESS OF FIT

31.1 FUNCTION

A Kolmogorov test of goodness of fit is used to determine whether an observed frequency distribution departs significantly from a hypothesized frequency distribution. Its use requires data on at least an ordinal scale, and the assumption is made that the data are continuously distributed. It is the appropriate alternative to the chi-square test of goodness of fit if the data are continuously distributed, and it may be used with much smaller samples than the chi-square test. However, the Kolmogorov test loses much of its power if the assumption of continuity is violated. (The presence of numerous tie scores is indicative of violation of the assumption of continuity.) The use of grouped data with the Kolmogorov statistic is highly questionable (and unlikely to lead to rejection of the null hypothesis), whereas the use of grouped data is mandatory with the chi-square statistic. The Kolmogorov procedure is simple to apply and usually requires less computational effort than the chi-square.

31.2 RATIONALE

Consider the situation in which a random sample is drawn from a continuously distributed population whose distribution is known. If the relative cumulative frequency of the population and that of the sample are plotted on a single graph, the *deviation* of the sample curve from the population curve will be independent of the population distribution. Thus, a single sampling distribution of this deviation suffices as a model for all samples of the same size, and it may be used for testing hypotheses about populations having any form of distribution. This sampling distribution has been determined for many different sample sizes and has been recorded in a form convenient for hypothesis testing in Table 13 of the Appendix.

The investigator need only determine the maximum deviation of the sample relative cumulative frequency distribution from the hypothesized relative cumulative frequency distribution. This maximum observed deviation is compared to the tabled value to determine if it is too large to be reasonably attributed to chance.

The assumption of continuity of distribution is one that is never met in practice—it is only approximated. As a result, the Kolmogorov test tends to be a conservative one, and in the practical research setting, the probability of rejecting a true null hypothesis is likely to be somewhat smaller than the level of significance set by the investigator. This problem is intensified by grouping the data, and the investigator is cautioned to avoid this practice.

31.3 PROCEDURE

(1) The application of a Kolmogorov goodness-of-fit test resembles that of a chi-square goodness-of-fit test. A theoretical distribution is either hypothesized or is known to exist in the completely random situation. The investigator must establish a hypothetical frequency distribution based upon whatever information he may have. This hypothesized distribution is translated into a relative cumulative frequency distribution with intervals that are directly comparable to sample intervals. Ordinarily, the author recommends that there be as many intervals as there are score levels.

(2) A random sample is drawn from the population under study, and its relative cumulative frequency distribution is determined.

(3) Each interval of the sample distribution is paired with the corresponding interval of the hypothesized distribution, and the difference between the two distributions is determined for each interval. Let D equal the maximum deviation.

(4) D is compared to the tabled value for the appropriate sample size and level of significance. A two-tailed test is appropriate (D may be either negative or positive), and this is what the tabled values provide. If the calculated value of D equals or exceeds the tabled value, the null hypothesis is rejected, the finding is significant, and the population distribution is determined to differ from the hypothesized distribution at the specified level of significance. If the calculated value is smaller than the tabled value, the hypothesis of no difference is retained.

31.4 EXAMPLE

The experiment that served to illustrate the use of a chi-square test of goodness of fit (see Section 28.4) will also be used to illustrate the use of a Kolmogorov test of goodness of fit. This was the situation in which students were awarded percentile rank scores on the basis of national norms, and the teacher sought to determine whether the students were distributed in the same fashion as the norm group. Unfortunately, the data were grouped in ten intervals when reported, and this may handicap the Kolmogorov test.

(1) H: The students are uniformly distributed.
     A: The students are not uniformly distributed.

(2) Kolmogorov goodness-of-fit test, $N = 60$.

(3) .05 level.
     R: $D \geqslant 0.18$.

(4) The value of D may be quite readily determined by arranging the data as shown in the following table. The sample data are from Section 28.4.

Percentile rank	10	20	30	40	50	60	70	80	90	
Hypothesis rcf	.10	.20	.30	.40	.50	.60	.70	.80	.90	1.0
Sample rcf	.15	.32	.38	.42	.47	.55	.57	.68	.82	1.0
Difference	.05	.12	.08	.02	.03	.05	.13	.12	.08	

$D = .13$

(5) Retain the null hypothesis. No significant difference between the two distributions has been determined to exist. The reader is reminded that the hypothesis was rejected when the chi-square test was used.

31.5 KOLMOGOROV TEST FOR GOODNESS OF FIT TO NORMAL

Let us suppose that the following collection of scores is a random sample from a much larger sample, and that the investigator desires to test the hypothesis that the parent population is normally distributed with mean of zero and standard deviation of one:

2.72,   2.01,   1.76,   1.29,   .67,   .10,   −.08,   −1.20,   −1.42,   −2.03

We will establish the hypothesized distribution with N equally probable intervals and present the data in tabular form as before:

z-score	−1.28	−.84	−.52	−.25	0	+.25	+.52	+.84	1.28
Hypothesis rcf	.10	.20	.30	.40	.50	.60	.70	.80	.90	1.0
Sample rcf	.20	.30	.30	.30	.40	.50	.50	.60	.60	1.0
Difference	.10	.10	0	.10	.10	.10	.20	.20	.30	

The value of D is .30, and this is nonsignificant at the .05 level $(R: \quad D \geqslant$ .41). Of course, a sample of size ten is hardly adequate to build a convincing case for normality of distribution.

The reader's attention is directed to the fact that there is no provision in the Kolmogorov test for estimating parameters from sample information. The hypothesis must completely specify the hypothesized distribution. In the case of the chi-square test for goodness of fit to normal, the investigator was privileged to estimate the population mean and standard deviation from sample information (resulting in the loss of two degrees of freedom).

31.6 DISCUSSION

The Kolmogorov maximum deviation test has been suggested as an alternative to the chi-square approximation of the multinomial for testing goodness of fit to a hypothesized distribution. The use of the chi-square statistic in goodness-of-fit tests usually necessitates rather large samples, while the Kolmogorov statistic may be used with samples of any size. The chi-square approximation of the multinomial may be used with either discrete or continuous data, while the Kolmogorov statistic requires data that are continuously distributed.

Ordinarily, when all of the assumptions are met and the statistician has a choice between two statistical tests, the more powerful of the two will be the one with the most rigorous assumptions. Thus, for a given sample size, one would anticipate that the Kolmogorov statistic would be superior to the chi-square when the data are continuously distributed. However, the Kolmogorov statistic suffers great loss of power if this assumption is violated, and this is quite likely to be the case in the practical research setting. Some experimentation on the part of the investigator may be necessary to determine which of the two tests is most appropriate for his data.

EXERCISES

31.1   Refer to Exercise 28.2. Substitute the Kolmogorov test for the chi-square.

31.2   An achievement test has been standardized with a large number of school children, and the scores are reported as percentile ranks based on the norm group. Use the Kolmogorov statistic to test the hypothesis that the following collection of scores is distributed in the same fashion as the norm group. Use the .05 level.

7	22	35	46	48
49	51	52	56	62
64	69	70	74	78
81	82	88	90	94

32

THE KOLMOGOROV-SMIRNOV TWO-SAMPLE TEST

32.1 FUNCTION

The Kolmogorov-Smirnov two-sample test is used to determine whether two populations are distributed in the same fashion. Its use requires data on at least an ordinal scale, and the assumption is made that the data are continuously distributed. It is the appropriate alternative to the chi-square test of independence for two samples if the data are continuously distributed, and it may be used with much smaller samples than the chi-square test. Like the Kolmogorov test of goodness of fit, the Kolmogorov-Smirnov two-sample test loses much of its power if the assumption of continuity is violated. If the data are grouped, the chi-square statistic will usually prove more powerful. The Kolmogorov-Smirnov procedure is simple to apply and usually requires less computational effort than the chi-square.

32.2 RATIONALE

The Kolmogorov-Smirnov two-sample test is an extension of the Kolmogorov statistic to the two-sample situation. If two independent random samples are drawn from the same continuously distributed parent population, and if their cumulative frequencies (or relative cumulative frequencies) are plotted on a single graph, the difference between the two curves will be independent of the distribution of the population. The sampling distribution of this difference has been determined for many different sample sizes and has been recorded in a form convenient

214

for hypothesis testing in Tables 14 and 15 of the Appendix. Table 14 is for use with small samples ($N \leqslant 40$) of equal size, and Table 15 is for use with large samples ($N \geqslant 20$) without the restriction of equal size.

The investigator need only determine the maximum difference between the cumulative frequencies of the two samples and compare it to the tabulated value to see if it is too large to be reasonably attributed to chance.

The assumption of continuity of distribution is one that is never met in practice—it is only approximated. As a result, the Kolmogorov-Smirnov test tends to be a conservative one, and in the practical research setting, the probability of rejecting a true null hypothesis is likely to be somewhat smaller than the level of significance set by the investigator.

32.3 PROCEDURE FOR SMALL SAMPLES OF EQUAL SIZE

(1) Two independent random samples of equal size are drawn by the investigator. These may be samples drawn from the same parent population and subjected to two different experimental treatments for determining the effects of the treatments, or they may be drawn from two different populations with the intent to determine whether the populations differ. Membership in one of the two samples constitutes the independent variable. The dependent variable is some continuously distributed criterion selected by the investigator.

(2) The cumulative frequency distributions of the two samples are established in such a fashion that the intervals of the two samples are directly comparable. The investigator is cautioned to use as many intervals as practical. The difference in the cumulative frequencies is determined for each interval, and the largest of these differences is designated K_D.

(3) K_D is compared to the tabled value for the appropriate sample size and level of significance. Table 14 of the Appendix is for samples of equal size up to $N = 40$, where N is the size of each sample. A one- or two-tailed test may be used. If the calculated value of K_D equals or exceeds the tabled value, the finding is significant, the null hypothesis is rejected, and the two distributions are determined to differ. The exact interpretation of a significant finding is dependent upon the nature of the criterion variable. If the calculated value is smaller than the tabled value, the null hypothesis is retained, and no significant difference in the two distributions has been established.

32.4 PROCEDURE FOR LARGE SAMPLES

(1) Two independent samples are drawn by the investigator. These may be samples drawn from the same parent population and subjected to two different experimental treatments, or they may be drawn from two different populations and compared on some criterion. In either event, a continuously distributed criterion measure must be available on each individual.

(2) The relative cumulative frequency distributions of the two samples are established in such a fashion that the intervals of the two samples are directly comparable. The investigator is cautioned to use as many intervals as practical. The difference in the relative cumulative frequencies is determined for each interval, and the largest of these differences is designated D.

(3) D is compared to the tabled value for the appropriate sample size and level of significance. Table 15 of the Appendix is for samples of size 20 or greater, and it is not necessary that the samples be of equal size. The table is two-tailed, but may be adapted for one-tailed use by being entered with twice the desired level of significance (use .10 in the table for a one-tailed test at the .05 level). If the calculated value of D equals or exceeds the tabled value, the finding is significant, and the null hypothesis is rejected. If the calculated value is smaller than the tabled value, the null hypothesis is retained.

32.5 EXAMPLE

The experiment that served to illustrate the use of the t-test for two independent samples (see Section 23.4) and the Mann-Whitney U-test (see Section 25.4) will also serve to illustrate the use of the Kolmogorov-Smirnov two-sample test. It should be noted that each of these three statistical analyses tests a different null hypothesis, but that each is a valid approach to the problem. To some extent, the selection of a statistical test is dependent upon the way in which the investigator sees the problem.

(1) H: No difference in the distributions associated with the two diets.
     A: Different distributions are associated with the two diets. The difference in the two might result from either greater gains or greater variability associated with one of the diets.

(2) Kolmogorov-Smirnov two-sample test, $n_a = n_b = 10$.

(3) .05 level, two-tailed test.
     R: $K_D \geqslant 7$.

(4) The raw scores will be used in establishing the two cumulative frequency distributions. The number of intervals will be set equal to the number of score levels.

Score	1	2	3	4	5	6	7	8	9	10	11
Sample A cf	1	2	3	5	7	7	7	8	10	10	10
Sample B cf	0	0	0	1	1	2	4	6	7	9	10
Difference	1	2	3	4	6	5	3	2	3	1	

$K_D = 6$

(5) Retain the null hypothesis; no significant difference in the two distributions was found. Both the t-test for independent samples and the Mann-Whitney U-test gave significant results with these data. The fact that the Kolmogorov-Smirnov test was unable to detect the difference between the two samples might be due to the fact that it tests a null hypothesis with a broader range of alternatives. Generally, a statistical test that may be rejected because of any one of several different kinds of departures from the sampling distribution will be less powerful than a statistical test that concentrates on a single alternative to the null hypothesis.

32.6 DISCUSSION

The Kolmogorov-Smirnov two-sample test has been suggested as an alternative to the chi-square test of independence for testing whether there are significant differences in the distributions of two samples. The use of the chi-square statistic in tests of independence usually necessitates rather large samples, while the Kolmogorov-Smirnov statistic may be used with samples of any size. The chi-square test may be used with either discrete or continuous data, while the Kolmogorov-Smirnov statistic requires data that are continuously distributed.

Under certain circumstances, the investigator may also regard the Kolmogorov-Smirnov two-sample test as an alternative to the t-test for two independent samples and the Mann-Whitney U-test.

The Kolmogorov-Smirnov two-sample test is a useful and versatile method of statistical analysis, but it suffers great loss in power if the assumption of continuity of distribution is violated. For this reason, it is not likely to be very useful where the data are grouped or where there are numerous tie scores. Some experimentation on the part of the investigator may be desirable to determine whether the Kolmogorov-Smirnov test provides an effective method of analyzing the data.

EXERCISES

32.1 Refer to Exercise 25.3. Substitute the Kolmogorov-Smirnov test for two independent samples for the Mann Whitney U-test.

32.2 Refer to Exercise 29.2. Substitute the Kolmogorov-Smirnov test for two independent samples for the chi-square test of independence.

THE FISHER EXACT PROBABILITY TEST

33.1 FUNCTION

The Fisher exact probability test is an extremely useful nonparametric test for use with two-by-two bivariate frequency tables. In contrast to the chi-square approximation of the multinomial, the Fisher test is based on exact probabilities and may be used with very small samples. Generally, the Fisher exact probability test is used to compare two independent samples on a dichotomous criterion. The two samples may be constituted by drawing them from a single population and subjecting them to experimentation, or by drawing them from two different populations for purposes of comparison.

33.2 RATIONALE

It is customary to let the letters A, B, C, and D represent the cell frequencies of a two-by-two bivariate frequency distribution:

A	B	$A + B$
C	D	$C + D$
$A + C$	$B + D$	

If the marginal frequencies are regarded as fixed, the probability of any particular arrangement of the cell frequencies may be determined from the hypergeometric distribution:

$$p = \frac{\binom{A + C}{A}\binom{B + D}{B}}{\binom{N}{A + B}} = \frac{(A + B)!\,(C + D)!\,(A + C)!\,(B + D)!}{N!\,A!\,B!\,C!\,D!}$$

In order to test a hypothesis with this distribution, it is necessary not only to determine the probability of occurrence of the particular frequency distribution observed, but also that of all more extreme distributions. The problem is essentially the same as that encountered in the use of the binomial probability distribution (see Section 18.4), in which the investigator sought to determine the probability that r was equal to or greater than some particular value. In the case of the hypergeometric distribution, the calculations get a little more involved; however, the distribution is tabled in a manner convenient for hypothesis testing in Table 16 of the Appendix.

Table 16 gives critical values of D (or C) corresponding to observed values of B (or A) when the marginal frequencies are fixed. If the calculated value of D (or C) is equal to or less than the tabled value, the finding is significant at the specified level. The table is one-tailed, but may be adapted to two-tailed tests by being entered with one-half the desired level of significance. In the case of two-tailed tests, the region of rejection must be established independently for each tail. Due to the discreteness of the data (fractional frequencies are not allowed), the tabled values are approximations, and any error is on the conservative side. Thus, the critical value listed at the .05 level might also be listed as the critical value for the .025 level, while the probability of rejecting a true hypothesis might be on the order of .022.

33.3 PROCEDURE

(1) All of the subjects in two independent random samples are assessed on a single dichotomous criterion. The data are arranged in a bivariate frequency table, and the cell frequencies are designated A, B, C, and D as outlined in the previous section.

(2) The marginal frequencies, $A + B$ and $C + D$, are determined, and the portion of Table 16 corresponding to these values is located.

(3) The table is entered for the observed value of B, and the observed value of D is compared to the critical value under the desired level of significance. If the observed value of B is not listed, the table is

entered for the observed value of A, and the observed value of C is compared to the critical value tabled under the desired level of significance.

(4) If the observed value of D (or C) is equal to or smaller than the tabulated value, the null hypothesis is rejected. If the observed value exceeds the tabled value, the null hypothesis is retained.

(5) The critical values listed in Table 16 are for a one-tailed test. A two-tailed test may be accomplished by entering the table with one-half the desired level of significance (use .025 in the table for a two-tailed test at the .05 level). If a two-tailed test is used, a separate determination of the region of rejection is required for each tail. The use of one- and two-tailed tests is illustrated in the following section.

33.4 EXAMPLES

Four examples will be given to illustrate the use of Table 16. In each case, a median test will be used so that the logic of one- and two-tailed tests can be demonstrated. The .05 level of significance will be used with all examples.

EXAMPLE I

	Group 1	Group 2	
Above median	$A = 7$	$B = 8$	$A + B = 15$
Below median	$C = 1$	$D = 14$	$C + D = 15$

One-tailed alternative: $Mdn_1 > Mdn_2$

$$R: \quad C \leqslant 1 \quad \text{(reject)}$$

One-tailed alternative: $Mdn_1 < Mdn_2$

$$R: \quad D \leqslant 2 \quad \text{(retain)}$$

Two-tailed alternative: $Mdn_1 \neq Mdn_2$

$$R: \quad C \leqslant 1, D \leqslant 1 \quad \text{(reject)}$$

EXAMPLE 2

	Group 1	Group 2	
Above median	$A = 6$	$B = 9$	$A + B = 15$
Below median	$C = 1$	$D = 14$	$C + D = 15$

One-tailed alternative: $Mdn_1 > Mdn_2$

$$R: \quad C \leqslant 1 \quad \text{(reject)}$$

One-tailed alternative: $Mdn_1 < Mdn_2$

$$R: \quad D \leqslant 3 \quad \text{(retain)}$$

Two-tailed alternative: $Mdn_1 \neq Mdn_2$

$$R: \quad C \leqslant 0, \; D \leqslant 2 \quad \text{(retain)}$$

EXAMPLE 3

	Group 1	Group 2	
Above median	$A = 5$	$B = 10$	$A + B = 15$
Below median	$C = 12$	$D = 3$	$C + D = 15$

One-tailed alternative: $Mdn_1 > Mdn_2$

$$R: \quad C \leqslant 0 \quad \text{(retain)}$$

One-tailed alternative: $Mdn_1 < Mdn_2$

$$R: \quad D \leqslant 4 \quad \text{(reject)}$$

Two-tailed alternative: $Mdn_1 \neq Mdn_2$

$$R: \quad C \leqslant 0, \; D \leqslant 3 \quad \text{(reject)}$$

EXAMPLE 4

	Group 1	Group 2	
Above median	$A = 4$	$B = 11$	$A + B = 15$
Below median	$C = 0$	$D = 15$	$C + D = 15$

One-tailed alternative: $Mdn_1 > Mdn_2$

$$R: \quad C \leqslant 0 \quad \text{(reject)}$$

One-tailed alternative: $Mdn_1 < Mdn_2$

$$R: \quad D \leqslant 5 \quad \text{(retain)}$$

Two-tailed alternative: $Mdn_1 \neq Mdn_2$

$$R: \quad D \leqslant 4 \quad \text{(retain)}$$

33.5 DISCUSSION

The Fisher exact probability test may be substituted for the chi-square test of independence when the data are cast into a two-by-two bivariate frequency table. Ordinarily, the Fisher test will be used only with those small samples provided for by Table 16, and the chi-square test will be used with larger samples. Occasionally, the Fisher test may be an appropriate alternative to the t-test for two independent samples or for the Mann-Whitney U-test.

EXERCISES

33.1 A random sample of 13 businesswomen are polled on whether they would favor a female president of the United States. They split, with 9 voting yes and 4 voting no. A random sample of 14 businessmen are polled on the same issue, and they split, with 3 voting yes and 11 voting no. Use the Fisher exact test to test the null hypothesis of no relationship between voting preference and sex of the respondent. Use a one-tailed test at the .05 level.

33.2 Repeat Exercise 33.1, substituting the chi-square test of independence for the Fisher exact test.

34

THE *F*-DISTRIBUTION

34.1   THE SAMPLING DISTRIBUTION OF THE *F*-RATIO

Consider the situation in which there are two normally distributed populations having the same variance. Let us draw a sample of size n_1 from the
first population and a sample of size n_2 from the second, calculate an
unbiased estimate of the common variance from each sample, then form
the ratio:

$$F = \frac{S_1{}^2}{S_2{}^2} \quad \text{where } S_1{}^2 = \frac{SS_1}{n_1 - 1} \quad \text{and} \quad S_2{}^2 = \frac{SS_2}{n_2 - 1}$$

If a very large number of samples are drawn in the fashion described
(all samples from the first population of size n_1 and all from the second of
size n_2), and the F-ratio is calculated for each pair of samples, the sampling
distribution of these ratios (which was derived by Sir Ronald Fisher) is
known as the F-distribution.

Like the t-distribution and the chi-square distribution, the F-distribu-
tion is a function of sample size, in this case, of the sizes of two samples.
There is a family of F-curves, each determined by two parameters, the
degrees of freedom associated with the numerator and the degrees of
freedom associated with the denominator. F-curves are positively skewed,
ranging from zero to infinity; those with large values of degrees of freedom
tend toward symmetry. The distribution of the F-ratio for representative
values of degrees of freedom is graphically illustrated in Fig. 34.1.

Because there are two values of degrees of freedom associated with
every F-ratio, F-tables tend to be bulky, and they are customarily tabled

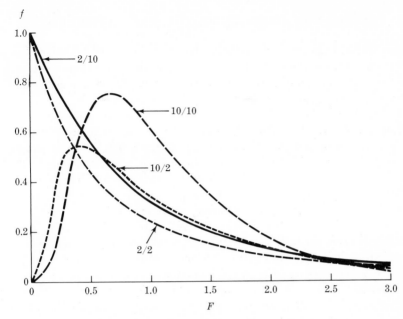

Figure 34.1   Representative *F*-Distributions.

only for the right-hand tail and for a few commonly used levels of significance. Table 17 in the Appendix is in two parts, the first giving critical
values of the *F*-ratio for the .05 level, the second for the .01 level. The
degrees of freedom for the numerator (designated by df_1) are given across
the top of the table and are used to determine which column is entered.
The degrees of freedom for the denominator (designated by df_2) are given
along the left side of the table and are used to determine which row is
entered. To use the tables, one must choose the page with the desired
level of significance, then find the point in the table where the column
and row for the appropriate degrees of freedom intersect. For example,
using the .01 level and 8, 20 degrees of freedom (degrees of freedom for the
numerator is always given first), the tabled entry is 3.56. An *F*-ratio equal
to or larger than the tabled value is significant.

As with the *t*-statistic, the table is not complete for all possible degrees
of freedom, and the investigator will occasionally find that the desired
degrees of freedom are not tabled. In this situation, he must discard
degrees of freedom, that is, enter the table with fewer degrees of freedom
than the problem merits. This procedure will necessitate a slightly larger
statistic to achieve rejection.

34.2 TESTING FOR HOMOGENEITY OF TWO VARIANCES

One of the assumptions underlying the t-test for two independent samples is that of homogeneity of variance—that the two samples are randomly drawn from populations having the same variance. The hypothesis of equal variances may be tested with the F-statistic.

Ordinarily, a two-tailed test is desired, and it is accomplished in the following fashion: The unbiased estimates of the variance are calculated for both samples, and the F-ratio is formed, the larger of these being in the numerator (the ratio should be equal to or larger than one). This ratio is compared to the tabled value for the appropriate degrees of freedom. The level of significance, however, is twice that specified by the table (the .01 table gives a two-tailed test at the .02 level).

If a one-tailed test is desired, the investigator should specify which variance is expected to be the larger of the two before the variance estimates are calculated. The ratio is formed using the estimate for this hypothesized larger variance in the numerator. Then the table is entered in the usual fashion (the .01 table gives a one-tailed test at the .01 level). Of course, if the F-ratio formed in this fashion is less than one, the finding is nonsignificant.

Unfortunately, the statistical test described is very sensitive to departures from normality, and it should not be used with small samples unless the assumption of normality is well established. This is unlikely to be the case in the practical research setting.

34.3 THE RELATIONSHIPS BETWEEN THE COMMON PARAMETRIC STATISTICS

It has been previously noted that the t-distribution with degrees of freedom equal to infinity is identical to the normal distribution. Mathematicians would say that the normal distribution is a special case of the t-distribution.

The F-distribution, with degrees of freedom of the numerator equal to one, is equal to the square of the t-distribution having the same degrees of freedom as the denominator of the F. It should be noted here that the one-tailed F is the square of the two-tailed t. The logic of this will become more apparent when the F-ratio is used to test hypotheses about means in Chap. 35. For example, the tabled t for 60 degrees of freedom is 2.00 at the .05 level. The tabled F for 1, 60 degrees of freedom is 4.00 at the .05 level. Thus, one may think of both the t-distribution and the normal distribution as special cases of the F-distribution.

The F-distribution, with the degrees of freedom of the denominator equal to infinity, may be multiplied by the degrees of freedom of the numerator to achieve a chi-square with the same degrees of freedom as the numerator of F. For example, the F-ratio with 10, ∞ degrees of freedom is 1.83 at the .05 level, and the chi-square with 10 degrees of freedom is 18.3 at the .05 level. Thus, the chi-square distribution is also a special case of the F-distribution.

34.4   REVIEW OF DOUBLE CLASSIFICATION NOTATION

At this point, the student is advised to review the use of the summation operator and statistical notation for data subject to double classification, as in the situation where the subjects are divided into two or more experimental samples (see Sections 6.3, 6.4, and 7.3).

If we let k represent the number of samples and n_j be the number of subjects in any sample, the total number of subjects in all samples is equal to:

$$N = \sum_{j=1}^{k} n_j$$

Similarly, if we let T_j represent the sum of the measures in any sample, the sum of the measures for all samples is equal to:

$$T = \sum_{j=1}^{k} T_j = \sum_{j=1}^{k} \sum_{i=1}^{n_j} X_{ij}$$

And, if we let M_j represent the mean of any sample, the general mean for all measures in all samples is equal to:

$$M = \frac{1}{N} \sum_{j=1}^{k} n_j M_j = \frac{1}{N} \sum_{j=1}^{k} T_j = \frac{T}{N}$$

34.5   PARTITIONING THE SUM OF SQUARES

Let SS_t represent the total sum of squares for k samples, the sum of the squared deviations from the general mean for all subjects in all samples:

$$SS_t = \sum_{j=1}^{k} \sum_{i=1}^{n_j} (X_{ij} - M)^2 = \sum_{j=1}^{k} \sum_{i=1}^{n_j} X_{ij}^2 - \frac{T^2}{N}$$

Let SS_w represent the sum of squares *within* groups—a measure of dispersion based upon variation within each sample around its sample mean:

$$SS_w = \sum_{j=1}^{k} \sum_{i=1}^{n_j} (X_{ij} - M_j)^2 = \sum_{j=1}^{k} \sum_{i=1}^{n_j} X_{ij}^2 - \sum_{j=1}^{n_j} \frac{T_j^2}{n_j}$$

And let SS_b represent the sum of squares *between* groups—a measure of dispersion that reflects variation of the sample means from the general mean:

$$SS_b = \sum_{j=1}^{k} n_j(M_j - M)^2 = \sum_{j=1}^{k} \frac{T_j^2}{n_j} - \frac{T^2}{N}$$

With a little effort and some elementary algebra, it can be demonstrated that:

$$SS_t = SS_w + SS_b$$

This exercise is known as "partitioning" the sum of squares, and it permits the investigator to analyze the contributions that various factors make toward the total variability of the individuals participating in a given experiment.

The degrees of freedom for SS_t is equal to $N - 1$, there being N terms in the expression and the general mean of the population being estimated from sample data. The degrees of freedom for SS_w is equal to $N - k$, there being N terms in the expression and k group means being estimated from sample data. Finally, the degrees of freedom for SS_b is equal to $k - 1$, there being k terms in the expression and the general mean being estimated from sample data. It is immediately apparent that the degrees of freedom were partitioned when the sum of squares was partitioned:

$$N - 1 = N - k + k - 1$$

34.6   THE ANALYSIS OF VARIANCE

The sum of squares within groups (SS_w) may be divided by its degrees of freedom ($N - k$) to obtain the mean square within (MS_w). If the data from which MS_w is calculated come from independent random samples of normally distributed and equally variable populations, the MS_w provides an unbiased estimate of the common population variance—an estimate that reflects variation within each sample around its sample mean.

The sum of squares between groups (SS_b) may be divided by its degrees of freedom ($k - 1$) to obtain the mean square between (MS_b). If the data from which MS_b is calculated come from independent random samples of normally distributed and equally variable populations with equal means, the MS_b also provides an unbiased estimate of the common population variance—an estimate that reflects variation of the sample means from the general mean.

If the conditions stated in the two previous paragraphs are met, and if MS_w and MS_b are calculated from the same data, then these data provide independent estimates of the same common population variance, and the following expression is distributed as F:

$$F = \frac{MS_b}{MS_w} \qquad \text{with } df = k - 1, \ N - k$$

If, however, the population means are unequal (but the other conditions are met), MS_b will tend to be larger than MS_w. The F-ratio thus calculated may be compared to the F-table to determine whether MS_b is significantly larger than MS_w. This procedure, known as the analysis of variance, provides a test of the hypothesis of equal means, which is mathematically equivalent to the t-test for independent samples in the two-sample situation. However, the analysis of variance is not limited to two samples; it may be used with as many samples as the investigator desires.

35

THE SIMPLE ANALYSIS
OF VARIANCE

35.1 FUNCTION

In all probability, the most useful experimental analysis considered in this text up to this point is the t-test for two independent samples. It provides what is generally the most powerful test of the hypothesis most frequently encountered in the behavioral sciences, that of equal means. However, the probability theory underlying the t-distribution is based upon the assumption that two (and only two) samples will be compared. If the number of comparisons is increased, the t-distribution no longer provides a valid test of the hypothesis of equal means. If the investigator uses the t-test to compare each pair of samples in a collection of three samples (a total of three comparisons instead of one), for example, the probability of rejecting a true hypothesis is much greater than the level of significance. The analysis of variance provides a statistical procedure that is appropriate for use with two *or more* samples. In the two-sample situation, it is mathematically equivalent to the t-test.

The simple analysis of variance is used for testing the hypothesis that two or more independent samples were drawn from populations having the same mean. The samples may be constituted by drawing independent random samples from a single population, subjecting them to experimentation, then comparing them on a single criterion variable. Or, the samples may be randomly drawn from different populations, then compared on a single criterion to determine whether the various populations differ with respect to this criterion.

230

35.2 RATIONALE

The major purpose of any experimental research is to describe the relationship between variation in the independent variable and variation in the dependent variable. The analysis of variance provides an extremely flexible procedure for determining the factors that influence the variation of the dependent variable. The total variation (as represented by the total sum of squares) is partitioned into its component parts in a fashion that permits the investigator to identify these as "sources" of variation.

In the simple analysis of variance, only two sources of variation are considered, that which is attributable to the independent variable (as represented by the sum of squares between groups), and that which is attributable to sampling error (as represented by the sum of squares within groups). These sums of squares are converted into variance estimates by dividing each by its respective degrees of freedom. These variance estimates are commonly referred to as the *mean square between* (MS_b) and the *mean square within* (MS_w). An F-ratio is formed by dividing the mean square between by the mean square within (the mean square within is the error term). The degrees of freedom for this F-ratio are those associated with the two mean squares ($k - 1$, $N - k$).

Four assumptions must be met for the F-ratio calculated in the simple analysis of variance to be distributed as the theoretical sampling distribution of F. The data used in the calculations must come from (1) independent random samples from (2) normally distributed and (3) equally variable populations with (4) equal means. If the first three of these assumptions are met, the magnitude of the mean square between is dependent upon the differences in the sample means. If the differences between the sample means are significant, the mean square between will be significantly larger than the mean square within. The F-ratio calculated from the sample data will be larger than the tabled value, and the hypothesis of equal means will be rejected.

To be completely accurate, the null hypothesis tested by the simple analysis of variance is that of no significant difference between the mean square between and the mean square within—with the one-tailed alternative that the mean square between is larger than the mean square within. A one-tailed test is achieved by always forming the F-ratio with the mean square between in the numerator and the mean square within in the denominator; this is compared to the right tail of the F-distribution as found in Table 17 in the Appendix. In the two-sample situation, this procedure yields precisely the same result as the t-test for two independent samples using the two-tailed alternative.

35.3   PROCEDURE

(1)   Two or more random samples are independently drawn from the population under study, and these are subjected to different experimental treatments. The analysis of variance may also be used to analyze existing differences between different populations.

(2)   At the conclusion of the experiment, a single criterion measure is administered to all subjects in all samples. The sums of squares between groups, within groups, and for total are calculated, along with their respective degrees of freedom. The sums of squares between groups and within groups are divided by their respective degrees of freedom to obtain the mean square between and the mean square within. The calculations needed to determine these values will be considered in great detail in the following section. It is customary to record these values in a summary table such as this:

Summary Table for the Analysis of Variance			
Source of variation	*df*	*Sums of squares*	*Mean squares*
Between groups			
Within groups			
Total			XXXXX

(3)   The F-ratio is calculated as follows:

$$F = \frac{MS_b}{MS_w} \quad \text{with } df = k - 1, \ N - k$$

(4)   The calculated F is compared to the tabled F at the desired level of significance and with the appropriate number of degrees of freedom. If the calculated statistic equals or exceeds the tabled value, the hypothesis is rejected, and a significant difference in the means of the various samples is determined to exist. Unfortunately, if there are more than two samples, this procedure does not indicate exactly which means are significantly different from each other. This is a complex problem, the solution to which will be sought in the next chapter. If the calculated statistic is smaller than the tabled value, the null hypothesis is retained, and no significant difference in the means has been found.

35.4 COMPUTATIONAL PROCEDURES

The sum of squares between groups has previously been identified as:

$$SS_b = \sum_{j=1}^{k} \frac{T_j^2}{n_j} - \frac{T^2}{N}$$

The sum of squares within groups has been identified as:

$$SS_w = \sum_{j=1}^{k} \sum_{i=1}^{n_j} X_{ij}^2 - \sum_{j=1}^{k} \frac{T_j^2}{n_j}$$

And the sum of squares for total has been identified as:

$$SS_t = \sum_{j=1}^{k} \sum_{i=1}^{n_j} X_{ij}^2 - \frac{T^2}{N}$$

The reader's attention is directed to the fact that some of the terms are held in common by these three formulas. As a matter of fact, only three terms need be evaluated to complete the calculations.

Consider the first term of the sum of squares within, which is also the first term of the sum of squares for total:

$$\sum_{j=1}^{k} \sum_{i=1}^{n_j} X_{ij}^2$$

To evaluate this term, one need only (a) square each score in the entire collection and (b) sum all of these squares.

Next, consider the second term of the sum of squares between, which is also the second term of the sum of squares for total:

$$\frac{T^2}{N}$$

To evaluate this term, one need only (a) sum all scores in the entire collection, (b) square this sum, and (c) divide by N.

Finally, consider the first term of the sum of squares between, which is also the second term of the sum of squares within:

$$\sum_{j=1}^{k} \frac{T_j^2}{n_j}$$

To evaluate this term, (a) sum the scores in each group, (b) square the sum for each group, (c) divide each of these squares by the number of subjects in the respective group, and (d) sum for all groups.

The degrees of freedom for these sums of squares are $k - 1$ for the sum of squares between groups, $N - k$ for the sum of squares within groups, and $N - 1$ for the sum of squares for total. Of course, k is the number of groups, and N is the total number of subjects in all groups.

The mean square between is simply the sum of squares between groups divided by $k - 1$, and the mean square within is the sum of squares within divided by $N - k$.

35.5   EXAMPLE

The students in a psychology class are randomly assigned to three different study groups. Each of the three groups is given a separate reading assignment to be completed within a period of two weeks. The first group is assigned reading in psychoanalytic theory, the second in cognitive-field theory, and the third in stimulus-response theory. At the end of the two-week period, a ten-item multiple-choice examination is administered to the class—an examination deliberately structured so that it favors stimulus-response theory. The results of the examination are reported below:

Group 1		Group 2		Group 3	
X	X^2	X	X^2	X	X^2
9	81	6	36	9	81
8	64	6	36	8	64
8	64	5	25	7	49
6	36	4	16	7	49
5	25	4	16	7	49
4	16	3	9	6	36
4	16	2	4	5	25
4	16	2	4		
48	318	32	146	49	353

(1) H: $\mu_1 = \mu_2 = \mu_3$ A: $\mu_1 \neq \mu_2 \neq \mu_3$.

(2)   Simple analysis of variance, $n_1 = 8$, $n_2 = 8$, $n_3 = 7$.

(3)   .05 level, $df = 2, 20$.
     R: $F \geqslant 3.49$.

(4) $\Sigma\Sigma X^2 = 318 + 146 + 353 = 817$

$$\sum \frac{T_j^2}{n_j} = \frac{48^2}{8} + \frac{32^2}{8} + \frac{49^2}{7} = 759$$

$$\frac{T^2}{N} = \frac{129^2}{23} = 723.52$$

$$SS_b = 759 - 723.52 = 35.48$$

$$MS_b = \frac{35.48}{2} = 17.74$$

$$SS_W = 817 - 759 = 58$$

$$MS_W = \frac{58}{20} = 2.90$$

$$SS_t = 817 - 723.52 = 93.48$$

Summary Table for the Analysis of Variance			
Source of variation	*df*	*Sums of squares*	*Mean squares*
Between groups	2	35.48	17.74
Within groups	20	58.00	2.90
Total	23	93.48	XXXXX

$$F = \frac{17.74}{2.90} = 6.12, \quad df = 2.20$$

(5)   Reject the null hypothesis. There is a significant difference in the means of the three samples. The mean of the first sample is 6, of the second sample is 4, and that of the third sample is 7. It is probably reasonable to conclude that the third group scored significantly higher than the second group, but nothing can be stated on the basis of the analysis of variance with respect to whether the third group scored significantly higher than the first, or whether the first scored higher than the second.

35.6   DISCUSSION

The simple analysis of variance provides one of the more popular statistical tests. It is a very powerful test that lends itself to a wide variety of research problems, and it is the basic unit of many more complex experimental designs.

The assumptions underlying the simple analysis of variance are essentially the same as those underlying the t-test for two independent samples. These assumptions are discussed in some detail in Section 23.5, and the conclusions there may be extended to the simple analysis of variance. These assumptions are that the samples are independent random samples from normally distributed and equally variable populations having the same means. When the hypothesis is rejected, we prefer to assume it is inequality of the means that has been violated. Fortunately, the analysis of variance is much more sensitive to violation of the assumption of equal means than to violation of the assumptions of normality or of homogeneous variances. Generally, when the investigator is working with samples of the same or nearly the same size, he may ignore these assumptions unless he has reason to believe that his measures deviate greatly from them. Of course, the assumption of normality may be ignored if the samples are of adequate size, due to the benefits of the central limit theorem.

There are nonparametric alternatives to the analysis of variance, but these are not nearly so popular as the nonparametric tests for two samples. As a general rule, nonparametric procedures are more readily adapted to simple analyses and are seldom extended to complex experimental designs. Only the chi-square tests of independence (especially the median test) of the nonparametric tests considered in this text are suitable for use with more than two samples.

Whenever the investigator uses the analysis of variance with more than two groups, there is a question as to which means are significantly larger than which other means. This is commonly known as the *problem of multiple comparisons*. This is a controversial topic among statisticians, and one for which there is at present no completely satisfactory solution. The reader should keep this in mind when he is tempted to regard the analysis of variance as an invitation to add experimental treatments to a research design. His attention is also directed to the fact that the addition of irrelevant treatments to a research design tends to reduce the power of the over-all test of the null hypothesis.

EXERCISES

35.1   Three groups of subjects selected at random participate in a psychological experiment. At the conclusion of the experiment, a test is administered, and the scores recorded are:

Group I: 0, 1, 1, 2
Group II: 1, 1, 2, 2
Group III: 3, 3, 4, 4

Use the analysis of variance to test the hypothesis of equal means at the .05 level.

35.2   A group of students are classified according to national heritage as belonging to Group A, B, C, or D. They are asked to respond to a rating scale that purports to assess the importance the individual places on formal education. A high score indicates that formal education is highly valued. The scores are recorded below:

$$\begin{aligned}
&\text{Group A:} \quad 0,\ 2,\ 5,\ 7 \\
&\text{Group B:} \quad 0,\ 1,\ 3,\ 5,\ 6 \\
&\text{Group C:} \quad 2,\ 4,\ 5,\ 5,\ 6,\ 9 \\
&\text{Group D:} \quad 1,\ 4,\ 6,\ 7,\ 8
\end{aligned}$$

Use the analysis of variance to test the hypothesis of equal means at the .05 level.

35.3   Refer to Exercise 23.1. Substitute the analysis of variance for the t-test for independent samples.

36

MULTIPLE COMPARISONS IN THE ANALYSIS OF VARIANCE

36.1 THE PROBLEM OF MULTIPLE COMPARISONS AMONG MEANS

The analysis of variance has been presented as a statistical procedure for making comparisons of means from two or more random samples. When the over-all hypothesis of equal means is rejected by the analysis of variance, this does not indicate that every sample mean differs significantly from every other sample mean. It is reasonable to assume, on the basis of rejection of the over-all null hypothesis, that at least one sample mean differs significantly from one other sample mean. If all of the samples are of the same size, the investigator is justified in concluding that the largest mean is significantly larger than the smallest mean. However, the contribution of any sample to a test of significance is a function of sample size, and if either the largest mean or the smallest mean comes from a sample with fewer subjects than one of the other samples, one cannot be sure that the null hypothesis is rejected due to the difference between these two extreme means. Even if the investigator can identify the difference between the largest mean and the smallest as a significant one, he will usually want to know whether there are other significant differences among the sample means. Generally, the rejection of the over-all hypothesis of equal means without identifying which means are significantly larger than which other means is not a very satisfactory conclusion to a piece of research.

Occasionally, it is suggested that the t-test for two independent samples may be used for testing the significance of the difference between pairs of means, if the test of the over-all hypothesis of equal means has

238

been rejected using the analysis of variance. The hypothesis tested by the *t*-test is that two random samples have been independently drawn from populations having the same mean. It is inappropriate for testing the significance of the difference between any two means from a collection of three or more samples. To illustrate the idea involved here, let us review some basic probability theory. If two coins are flipped simultaneously, the probability of getting two heads is 1/4. If three coins are flipped simultaneously, the probability of two or more heads is 1/2. If four coins are flipped simultaneously, the probability of getting two or more heads is 11/16. The same idea applies in sampling for purposes of comparing means; one cannot use a probability model based upon the difference between two random samples to test hypotheses about any two samples from a larger collection of random samples.

The problem of multiple comparisons in the analysis of variance is a controversial topic among statisticians, and it is a problem that has no simple solution. Several procedures have been developed for making multiple comparisons, but some of these tend to yield significant findings when the test of the over-all hypothesis provided by the analysis of variance is nonsignificant. In view of the fact that it has not been established that any of these tests is more powerful than the test provided by the analysis of variance, their validity is subject to question. Some alternative procedures for making multiple comparisons are conservative and tend to yield nonsignificant findings when the test of the over-all hypothesis is significant. Such procedures are mathematically defensible but not very appealing. The Scheffé procedure is one of these, and its use is recommended when the investigator wishes to make all possible comparisons between pairs of means.

Occasionally, the investigator is prepared to specify in advance of the drawing of samples that he is interested only in comparing one sample to all other samples. This will usually be thought of as comparing one control group to two or more experimental groups. When the investigator is willing to place this restriction upon the comparisons to be made, the Dunnett procedure is recommended, and it will prove more powerful than the Scheffé. Like the analysis of variance, both the Scheffé and the Dunnett tests are mathematically equivalent to the *t*-test when only two samples are drawn.

36.2 THE SCHEFFÉ TEST FOR ALL POSSIBLE COMPARISONS

The Scheffé procedure for testing any and all possible comparisons between means has the important property that the probability of a Type I error for any comparison does not exceed the level of significance specified in the analysis of variance for the over-all hypothesis. However, the probability of a Type I error for a given comparison may be considerably smaller than the level of significance set by the investigator. Like the analysis of variance, the Scheffé procedure is quite insensitive to departures from normality and homogeneity of the variances. The test statistic is quite simply calculated for any pair of means, and it is referred to the same region of rejection as that specified for the test of the over-all hypothesis of equal means. The formula is as follows:

$$F = \frac{(M_1 - M_2)^2}{MS_w \left(\frac{1}{n_1} + \frac{1}{n_2}\right)(k - 1)} \qquad \text{with } df = k - 1, \ N - k$$

The problem used to illustrate the simple analysis of variance (Section 35.5) will also be used to illustrate the Scheffé technique. The region of rejection was determined to be F equal to or greater than 3.49 in the previous illustration.

Comparing the means for Groups 1 and 2, we obtain:

$$F = \frac{(6 - 4)^2}{2.9(\frac{1}{8} + \frac{1}{8})(2)} = 2.76 \qquad \text{(retain)}$$

Next, comparing the means for Groups 1 and 3 we obtain:

$$F = \frac{(7 - 6)^2}{2.9(\frac{1}{8} + \frac{1}{7})(2)} = 0.64 \qquad \text{(retain)}$$

Finally, comparing the means for Groups 2 and 3, we obtain:

$$F = \frac{(7 - 4)^2}{2.9(\frac{1}{8} + \frac{1}{7})(2)} = 5.79 \qquad \text{(reject)}$$

The conclusion is that Group 3 scored significantly higher than Group 2, but the differences between Group 1 and Group 2 and between Group 1 and Group 3 are both nonsignificant.

The Scheffé procedure is not limited to simple comparisons of pairs of means, but may be used to study more complex relationships. Suppose the investigator in the previous example wished to compare the mean of Group 3 to the mean of Groups 1 and 2 combined:

$$M_{1+2} = \frac{n_1 M_1 + n_2 M_2}{n_1 + n_2} = 5$$

$$F = \frac{(M_3 - M_{1+2})^2}{MS_w \left(\frac{1}{n_1 + n_2} + \frac{1}{n_3}\right)(k - 1)} = 3.37 \qquad \text{(retain)}$$

When the various samples are all of equal size, the Scheffé procedure may be used to arrive at a critical difference between the means, thereby saving some computational effort:

$$d = \sqrt{\frac{2(k-1)(\text{tabled } F)(MS_w)}{n}}$$

The tabled F in this case is the region of rejection selected earlier, and the value of d is the minimum difference between any two means necessary for a significant finding.

Let us pretend, for purposes of illustration, that all of the samples in the illustration above had eight subjects. The critical difference is:

$$d = \sqrt{\frac{2(2)(3.49)(2.9)}{8}} = 2.25$$

The difference between Means 1 and 2 in the example is two (which is nonsignificant), the difference between Means 1 and 3 is one (which is nonsignificant), and the difference between Means 2 and 3 is three (which is significant). This is consistent with the earlier conclusions, even though the size of one sample was juggled for purposes of illustration.

Unfortunately, it is not at all uncommon to follow a significant finding by the analysis of variance with the Scheffé procedure and find that no two means differ significantly. This may be attributed to the fact that the analysis of variance provides a more powerful test of the hypothesis of equal means. When this occurs, if the various samples are of equal (or nearly equal) size, and the analysis of variance yields a significant finding, it is reasonable to conclude that the largest mean is significantly larger than the smallest mean, even though the Scheffé test was unable to detect this difference. Of course, the investigator may wonder whether there are other significant differences that are not detected by the Scheffé test.

36.3 THE DUNNETT TEST FOR MULTIPLE COMPARISONS TO A CONTROL GROUP

Dunnett has derived the sampling distribution for a t statistic for use in comparing a control group to one or more experimental groups. Dunnett's t is calculated as follows:

$$t = \frac{M_1 - M_2}{\sqrt{MS_w \left(\dfrac{1}{n_1} + \dfrac{1}{n_2}\right)}} \qquad \text{with } df = N - k$$

Provision is made for both one- and two-tailed tests in the distribution of Dunnett's t in Table 18 of the Appendix.

The problem used to illustrate the simple analysis of variance (Section 35.5) and the Scheffé procedure (Section 36.2) will also serve to illustrate the Dunnett test. A review of the experiment reveals that it is consistent to undertake a one-tailed test with the alternative that Group 3 will score higher than either Group 1 or Group 2. Referring to Table 18, we find that the region of rejection (.01 level, one-tailed, two experimental groups, $df = 20$) is determined to be t equal to or greater than 2.81.

Comparing Group 3 to Group 1, we obtain:

$$t = \frac{7 - 6}{\sqrt{2.6(\frac{1}{8} + \frac{1}{7})}} = 1.2 \quad \text{(retain)}$$

Comparing Group 3 to Group 2, we obtain:

$$t = \frac{7 - 4}{\sqrt{2.6(\frac{1}{8} + \frac{1}{7})}} = 3.5 \quad \text{(reject)}$$

Like the Scheffé procedure, the Dunnett procedure is conservative, and the probability of a Type I error for any comparison does not exceed the level of significance specified in the analysis of variance for the over-all hypothesis.

EXERCISES

36.1 Refer to Exercise 35.1. Use the Scheffé procedure to make all possible comparisons between pairs of means.

36.2 Refer to Exercise 35.1. Use the Dunnett procedure to make a one-tailed comparison of Group I to all other groups.

37

TWO-DIMENSIONAL
ANALYSIS OF VARIANCE

37.1 FUNCTION

Often, the most valuable research conducted by the behavioral scientist is that in which there are two or more independent variables (in addition to the dependent variable). In this chapter and those that follow, three different approaches to research analysis involving two or more independent variables will be considered: (1) two-dimensional analysis of variance, (2) analysis of covariance, and (3) multiple regression. Traditionally, two-dimensional analysis of variance has been associated with experimental designs employing two discrete independent variables, the analysis of covariance with designs using one discrete and one or more continuous variables, and multiple regression with two or more continuous independent variables. Of course, the student is now aware that continuous data may be treated as discrete, and that it is often useful to transform discrete data into binary variables and to complete the analysis with procedures ordinarily reserved for continuously distributed data. For these reasons, the various techniques considered in these chapters will often be interchangeable, and the selection of the statistical analysis will depend to some extent on how the investigator perceives his data and his objectives.

Two-dimensional analysis of variance is an extension of simple analysis of variance to the situation in which there are two independent variables. It is customary to organize the data into a double-entry table, with one of the independent variables cast into columns and the other into rows. Scores on the dependent variable are entered in the cells of the table. Consider, for example, the following two-by-two table:

	Drug	No drug
Electroshock	$X_{111}, X_{112} \ldots X_{11k}$	$X_{121}, X_{122} \ldots X_{12k}$
No shock	$X_{211}, X_{212} \ldots X_{21k}$	$X_{221}, X_{222} \ldots X_{22k}$

The table represents a two-dimensional experimental design in which a psychiatrist is studying concomitantly the effects of a certain drug and electroshock therapy. Patients in cell 11 receive both the drug and shock treatments, those in cell 12 only the shock, those in cell 21 only the drug, and those in cell 22 neither the drug nor the shock treatment. Notice that all patients in column 1 receive the drug, while those in column 2 do not; and that all patients in row 1 receive the shock therapy, while those in row 2 do not. The investigator is enabled by this design to carry on simultaneously two experiments with the same group of subjects. In addition to the economies effected by carrying on two experiments with the same subjects at the same time, the investigator is enabled to study possible interaction between the row and column variables. Interaction is a treatment effect that is dependent upon the concomitant influence of two independent variables. In the example, a significant interaction would suggest that (1) the effects of the shock-treatment variable are not the same for those patients who received the drug as for those who did not, and that (2) the effects of the drug-treatment variable are not the same for those patients who received the shock treatment as for those who did not. A nonsignificant interaction would suggest that the drug and shock treatment effects are independent of each other.

Two-dimensional analysis of variance may also be used to improve upon the simple analysis of variance through the use of stratified random samples. The educational researcher, for example, might find it desirable to stratify a population of school children, dividing it into three groups, those of high IQ, those of average IQ, and those of low IQ. Then independent random samples might be drawn from each of these three levels and cast into a two-by-three table for purposes of studying the effects of two methods of instruction:

	Method A	*Method B*
High IQ		
Average IQ		
Low IQ		

From a computational standpoint, the analysis is the same as that in which the investigator is interested in studying the effects of two treatment variables; however, the interpretation of the results may be quite different. Assuming that the stratification variables (IQ in the example) are related to the criterion, the precision of the experiment is improved through the use of matched samples, and the value of the mean square within will be smaller. The investigator also has the option to draw separate conclusions from each level of this variable if he so desires.

37.2 RATIONALE

Data organized in double-entry tables of the sort described in the previous section may be approached in three different fashions with simple analysis of variance. First, the existence of the row variable may be ignored and the sum of squares for total (SS_t) partitioned into the sum of squares between columns (SS_c) and the sum of squares within columns. Second, the existence of the column variable may be ignored and the sum of squares for total (SS_t) partitioned into the sum of squares between rows (SS_r) and the sum of squares within rows. Finally, the cells in the table may be treated as distinct groups and the sum of squares for total (SS_t) partitioned into the sum of squares between cells (SS_{cells}) and the sum of squares within cells (SS_w), which we shall refer to simply as the *sum of squares within*. The student should note that the sum of squares for total is identical in the three analyses described above, and that the procedures suggested are identical to those presented in Chap. 37 on the simple analysis of variance.

Suppose, however, that upon completing the calculations described above, the investigator determines that the sum of squares for cells is substantially larger than the sum of squares for columns plus the sum of squares for rows. In other words, the variability from cell to cell cannot be entirely accounted for by the effects of the row and column variables. This suggests the possibility of a third treatment effect, one that is due to interaction between the row and column variables. This is referred to as the *sum of squares for interaction* (SS_{rc}), and it is equal to the sum of squares for cells minus the sum of squares for columns and the sum of squares for rows:

$$SS_{rc} = SS_{cells} - SS_c - SS_r$$

The student may find it useful to think of this as partitioning the sum of squares for cells into its component parts.

The commonly accepted approach to two-dimensional analysis of variance consists of partitioning the sum of squares for total into four parts: (1) sum of squares for columns, (2) sum of squares for rows, (3) sum of squares for interaction, and (4) sum of squares within. As in the simple analysis of variance, each of these sums of squares will be converted into a mean square by dividing each by its respective degrees of freedom. The actual computations will be treated in detail in a later section, where the student may demonstrate for himself that the total sum of squares has in fact been partitioned into four parts:

$$SS_t = SS_c + SS_r + SS_{rc} + SS_w$$

In similar fashion, one may demonstrate that the total number of degrees of freedom has been partitioned:

$$(N - 1) = (c - 1) + (r - 1) + (c - 1)(r - 1) + (N - rc)$$

where c is the number of columns and r the number of rows.

The test for significance of interaction is accomplished by forming the ratio:

$$F = \frac{MS_{rc}}{MS_w} \quad \text{with } df = (r - 1)(c - 1), \quad (N - rc)$$

Similarly, the test for significance of the column variable is accomplished by forming the ratio:

$$F = \frac{MS_c}{MS_w} \quad \text{with } df = (c - 1), \quad (N - rc)$$

This F-ratio provides a test of the hypothesis of equal column means, ignoring the existence of the rows. Such a test is sometimes referred to as a *test of main effects for columns*.

Finally, the test for significance of the row variable is accomplished by forming the ratio:

$$F = \frac{MS_r}{MS_w} \quad \text{with } df = (r - 1), \quad (N - rc)$$

This F-ratio provides a test of the hypothesis of equal row means, ignoring the existence of the columns. Such a test is sometimes referred to as a *test of main effects for rows*.

Ordinarily, a significant interaction effect suggests that the tests of main effects for rows and columns outlined above are inappropriate. The presence of a significant interaction effect indicates that the row effects are not constant from column to column, and that the column effects are not constant from row to row. The accepted procedure in this situation is to break the total design into parts, performing a simple analysis of variance for columns separately for each row in the table, and performing a simple analysis of variance for rows separately for each column in the table. These tests are commonly referred to as *tests of simple effects for rows and columns.*

Underlying the derivation of the two-dimensional analysis of variance presented in this chapter is a requirement of proportionality of cell frequencies from row to row and column to column. For a two-by-three double-entry table, this requirement may be stated algebraically by:

$$\frac{n_{11}}{n_{21}} = \frac{n_{12}}{n_{22}} = \frac{n_{13}}{n_{23}}$$

Often, this will mean that the investigator will choose to use random samples of the same size for the various cells; however, if he has a reason for doing so, the investigator might choose samples for a two-by-three design of unequal sizes such that:

$$\frac{10}{20} = \frac{15}{30} = \frac{20}{40}$$

Of course, this type of restriction precludes the use of available groups of odd sizes such as those commonly encountered in classroom research.

37.3 PROCEDURE

(1) A double-entry table is constructed in the fashion described in Section 37.1, random samples are independently drawn from the population under study (one for each cell in the table), and these are subjected to experimentation. The analysis may also be used to analyze existing differences between different populations where these populations logically fit a double-entry table. The cell frequencies must meet the requirement of proportionality outlined in Section 37.2.

(2) At the conclusion of the experiment, a single criterion measure is administered to all subjects in all samples. A summary table is constructed, and the various degrees of freedom, sums of squares, and

mean squares needed to complete the analysis are calculated. The following summary table serves as an example:

Summary Table for Two-dimensional Anaylsis of Variance			
Source of variation	*df*	*Sums of squares*	*Mean squares*
(Between cells)	XX		XXXX
Between columns			
Between rows			
Interaction			
Within cells			
Total			XXXX

The student's attention is directed to the fact that while the sum of squares between cells does not enter into the final analysis, it is a useful intermediate step in the calculation of the sum of squares for interaction.

(3) The significance of interaction is tested first by forming the ratio of MS_{rc} to MS_w. This is compared to the tabled F-ratio with the appropriate level of significance and degrees of freedom. If the interaction is significant, the effects of the row variable are not the same for each column, and the effects of the column variable are not the same for each row. If the interaction is nonsignificant, the row and column variables may be considered independently.

(4) If the interaction is significant, the analysis will be completed for the simple effects of the row and column variables. In the case of a two-by-three design, for example, there will be three simple analyses of variance with two groups each and two simple analyses with three groups each. If the interaction is nonsignificant, the main effects for columns may be tested by the ratio of MS_c to MS_w and the main effects for rows by the ratio of MS_r to MS_w. These, of course, are tests of the significance of difference between the column and row means, respectively.

37.4 COMPUTATIONAL PROCEDURES

The sum of squares between columns may be calculated from:

$$SS_c = \sum_{j=1}^{c} n_j(M_j - M)^2 = \sum_{j=1}^{c} \frac{T_j^2}{n_j} - \frac{T^2}{N}$$

where n_j is the number of subjects, M_j the mean, and T_j the sum of the scores in the jth column.

Similarly, the sum of squares between rows may be calculated from:

$$SS_r = \sum_{i=1}^{r} n_i(M_i - M)^2 = \sum_{i=1}^{r} \frac{T_i^2}{n_i} - \frac{T^2}{N}$$

where n_i is the number of subjects, M_i the mean, and T_i the sum of the scores in the ith row.

And the sum of squares between cells may be calculated from:

$$SS_{\text{cells}} = \sum_{i=1}^{r} \sum_{j=1}^{c} n_{ij}(M_{ij} - M)^2 = \sum_{i=1}^{r} \sum_{j=1}^{c} \frac{T_{ij}^2}{n_{ij}} - \frac{T^2}{N}$$

where n_{ij} is the number of subjects, M_{ij} the mean, and T_{ij} the sum of the scores in the ijth cell.

Both the sum of squares for interaction and the sum of squares within cells will be calculated as residuals:

$$SS_{rc} = SS_{\text{cells}} - SS_c - SS_r$$
$$SS_w = SS_t - SS_{\text{cells}}$$

Finally, the sum of squares for total may be calculated in the usual fashion:

$$SS_t = \sum_{i=1}^{r} \sum_{j=1}^{c} \sum_{k=1}^{n_{ij}} X_{ijk}^2 - \frac{T^2}{N}$$

The reader's attention is directed to the fact that some of the terms are held in common by these formulas. As a matter of fact, only five terms need be evaluated to complete the calculations.

Consider the first term of the sum of squares for total:

$$\sum_{i=1}^{r} \sum_{j=1}^{c} \sum_{k=1}^{n_{ij}} X_{ijk}^2$$

To evaluate this term, one need only (a) square each score in the entire collection and (b) sum all of these squares.

Next, consider the second term of the sum of squares between columns, between rows, between cells, and for total:

$$\frac{T^2}{N}$$

To evaluate this term, one need only (a) sum all scores in the entire collection, (b) square this sum, and (c) divide by N.

The first term of the sum of squares between columns is:

$$\sum_{j=1}^{c} \frac{T_j^2}{n_j}$$

To evaluate this term, (a) sum all scores for each column, (b) square each sum, (c) divide each of the squares by the number of subjects in the respective column, and (d) sum for all columns.

The first term of the sum of squares between rows is:

$$\sum_{i=1}^{r} \frac{T_i^2}{n_i}$$

To evaluate this term, (a) sum all scores for each row, (b) square each sum, (c) divide each of the squares by the number of subjects in the respective row, and (d) sum for all rows.

Finally, the first term of the sum of squares between cells is:

$$\sum_{i=1}^{r} \sum_{j=1}^{c} \frac{T_{ij}^2}{n_{ij}}$$

To evaluate this term, (a) sum all scores for each cell, (b) square each sum, (c) divide each of the squares by the number of subjects in the respective cell, and (d) sum for all cells.

The degrees of freedom for the various sums of squares are $(c - 1)$ for between columns, $(r - 1)$ for between rows, $(c - 1)(r - 1)$ for interaction, $(N - rc)$ for within, and $(N - 1)$ for total. Of course, c is the number of columns, r the number of rows, and N the number of subjects in the entire collection.

As always, a mean square is simply a sum of squares divided by its degrees of freedom.

37.5 EXAMPLE

The psychiatric experiment described in the earlier sections of the chapter will serve to illustrate the use of two-dimensional analysis of variance. Suppose that four independent random samples of size four are drawn and assigned to the four treatment combinations. At the conclusion of the experiment, a criterion of some sort is administered to the patients, and their scores are recorded in the double-entry table as follows:

	Drug	No drug
Electroshock	2, 3, 3, 4	1, 2, 2, 3
No shock	0, 1, 2, 3	0, 1, 1, 2

(1) $H1$: There is no interaction between the row and column variables.
$H2$: The column means are equal.
$H3$: The row means are equal.

(2) Two-dimensional analysis of variance, $n_{11} = n_{21} = n_{12} = n_{22} = 4$.

(3) .05 level, R: $F \geqslant 4.75$, $df = 1$, 12 for all three hypotheses (the region of rejection will not always be the same for all three hypotheses in two-dimensional analysis of variance).

(4) $\Sigma\Sigma\Sigma X_{ijk}^2 = 76.00$

$$\frac{T^2}{N} = \frac{30^2}{16} = 56.25$$

$$\sum \sum \frac{T_{ij}^2}{n_{ij}} = \frac{12^2}{4} + \frac{8^2}{4} + \frac{6^2}{4} + \frac{4^2}{4} = 65.00$$

$$\sum \frac{T_j^2}{n_j} = \frac{18^2}{8} + \frac{12^2}{8} = 58.50$$

$$\sum \frac{T_i^2}{n_i} = \frac{20^2}{8} + \frac{10^2}{8} = 62.50$$

$SS_{\text{cells}} = 65.00 - 56.25 = 8.75$
$SS_c = 58.50 - 56.25 = 2.25$
$SS_r = 62.50 - 56.25 = 6.25$
$SS_{rc} = 8.75 - (6.25 + 2.25) = .25$
$SS_w = 19.75 - 8.75 = 11.00$
$SS_t = 76.00 - 56.25 = 19.75$

Summary Table for Two-dimensional Analysis of Variance			
Source of variation	*df*	*Sums of squares*	*Mean squares*
(Between cells)	XX	(8.75)	XXXX
Between columns	1	2.25	2.25
Between rows	1	6.25	6.25
Interaction	1	.25	.25
Within cells	12	11.00	.917
Total	15	19.75	XXXX

$$\text{Interaction:} \quad F = \frac{0.25}{0.917} = .27 \qquad \text{(retain)}$$

$$\text{Column means:} \quad F = \frac{2.45}{0.917} = 2.67 \qquad \text{(retain)}$$

$$\text{Row means:} \quad F = \frac{6.25}{0.917} = 6.82 \qquad \text{(reject)}$$

(5)   There is no significant interaction between the drug and shock treatments. There is no significant difference between the column means; apparently the drug is ineffective. There is a significant difference between the row means with those patients receiving electroshock therapy scoring higher on the criterion.

37.6   DISCUSSION

Two-dimensional analysis of variance is an extension of simple analysis of variance to the situation in which the investigator is interested in studying the effects of two independent variables on a single criterion.

In addition to the assumptions underlying the simple analysis of variance, two-dimensional designs also have the requirement of proportional cell frequencies from column to column and from row to row. This requirement is a severe restriction in much behavioral research and may preclude the use of this design in many research situations. Although a number of solutions to this problem have been offered, they are beyond the scope of this book.

The ideas used in developing two-dimensional analysis in this chapter may be extended to repeated measurements designs plus three-dimensional and higher order designs; however, these are also beyond the scope of this book.

EXERCISES

37.1 A group of freshman college students is classified according to sex and according to whether they graduated from an urban or rural high school. Their grades (as recorded on a four-point scale) in a first course in public speaking are recorded in the double-entry table below:

	Urban H. S.	Rural H. S.
Males	0, 1, 2, 2, 3	0, 0, 1, 2, 3
Females	1, 2, 3, 4, 4	1, 2, 2, 2, 3

Construct a summary table for two-dimensional analysis of variance, and:

(a) Test the hypothesis of no interaction between the row and column variables.

(b) Test the main effects for the column variable. Is there a significant difference in the performance of the urban and rural students?

(c) Test the main effects for the row variable. Is there a significant difference in the performance of the male and female students?

38

THE ANALYSIS
OF COVARIANCE

38.1 FUNCTION

The success of an experiment and the ability to detect significant differences in the criterion variable are often determined by the ability of the investigator to control one or more variables that influence the criterion. In Chaps. 24 and 26, the possibility of controlling variables through the use of matched pairs or repeated measurements designs was discussed; in Chap. 37, the possibility of controlling variables through the use of stratified samples was considered. Each of these techniques uses experimental procedures to control the influence of these variables on the criterion, and each introduces special problems with respect to the conduct of the experiment. Random samples stratified into two or three levels on some variable related to the criterion may be used with two-dimensional analysis of variance to grossly control the influence of the stratification variable; however, the requirement of proportionality of cell frequencies prohibits the use of this design in some research. The use of repeated measurements designs permits simultaneous control of many variables, but this design is inapplicable in the vast majority of behavioral experiments due to the carry-over from treatment to treatment and from measure to measure. Occasionally, the use of matched pairs can be substituted to achieve the same end, but with less precision. The analysis of covariance is a blending of regression and the analysis of variance, which permits *statistical* rather than *experimental* control of variables. The result is equivalent to matching the various experimental groups with respect to the variable or variables being controlled.

The use of analysis of covariance ordinarily involves a pretest (the

254

variable to be controlled) and a posttest (the criterion) that are known to be correlated. IQ scores, achievement test scores, and previous course grades are often used as pretest measures in educational research. In some circumstances, it may be completely appropriate to use the same instrument for both pretest and posttest. Occasionally, the variable to be controlled is a stable trait that is unlikely to be influenced by experimental manipulation (sex or IQ, for example), and the actual recording of the measure may take place during or following the experiment.

Whenever two measures are correlated, one can be used to predict scores on the other; to the extent that performance on the posttest can be predicted from performance on the pretest, this performance cannot be attributed to the experimental activities. The analysis of covariance consists essentially of determining that a proportion of the variance of the criterion existed prior to the experiment, and this proportion is eliminated from the final analysis. It should be immediately apparent that two substantial benefits accrue from such a procedure: (1) any variable that influences the variation of the criterion variable may be controlled, and (2) the error variance in the analysis is substantially reduced.

Like the simple analysis of variance, the analysis of covariance may be used with any number of samples. It may be substituted for the simple analysis of variance whenever a suitable pretest is available. In fact, much educational research can be improved by introducing the analysis of covariance *after* the experiment has been completed. One need only find in the students' files some measure such as IQ, achievement test score, or previous grade that can be paired with the criterion score for the analysis.

38.2 RATIONALE

The essential task in statistical analysis is to account for the variability of the criterion. Variance has been defined as the mean of the squared deviations from the mean, and it may be calculated from the formula:

$$\sigma^2 = \frac{SS}{N}$$

In the analysis of variance, the total sum of squares is partitioned into its component parts in an attempt to identify the sources of variability of the criterion. The r-square interpretation of correlation is a similar procedure.

The analysis of covariance is concerned with concomitant variation in the criterion variable and a variable whose relationship to the criterion is to be controlled. Covariance may be defined as the mean of the products (for paired variables) of the deviations from the mean, and it may be calculated from the formula:

$$\text{cov} = \frac{SP}{N}$$

It should be noted that covariance may be either negative or positive; variance, of course, is always a positive quantity.

The analysis of covariance involves the partitioning of the total sum of products in a fashion similar to the partitioning of the total sum of squares in the analysis of variance. The total sum of products may be calculated from:

$$SP_t = \sum_{j=1}^{k} \sum_{i=1}^{n_j} X_{ij} Y_{ij} - \frac{T_x T_y}{N}$$

where T_x and T_y are the sums of the respective X- and Y-measures for all groups.

The sum of products between groups may be calculated from:

$$SP_b = \sum_{j=1}^{k} \frac{T_{xj} T_{yj}}{n_j} - \frac{T_x T_y}{N}$$

where T_{xj} and T_{yj} are the sums of the X- and Y-measures for the jth group.

The sum of products within groups may be calculated as a residual:

$$SP_w = SP_t - SP_b$$

We will follow the custom of letting X represent the predictor (controlled) variable and Y the predicted (criterion) variable. The student is reminded that there is a third variable of great importance, the independent variable whose effects on the criterion are to be studied, this variable being represented by membership in one of the experimental groups. If we introduce the assumption that the slope of the regression line (predicting Y from X) is the same for the various experimental groups, the best estimate of this common slope may be obtained from:

$$b_w = \frac{SP_w}{SS_{wx}}$$

where SS_{wx} is the sum of squares within for the X-measures.

The predicted criterion score (in deviation form) for any individual is:

$$\hat{y} = b_w x$$

This predicted score is the portion of the criterion measure that may be determined from a knowledge of the variable to be controlled. It may be subtracted from the criterion score to obtain an adjusted criterion score:

$$y' = y - b_w x \quad \text{or} \quad Y' = Y - b_w x$$

It is these adjusted scores upon which the final analysis is based. Statistical inferences are drawn with respect to adjusted group means, which are calculated in the following manner:

$$\bar{Y}'_j = \bar{Y}_j - b_w(\bar{X}_j - M_x)$$

where M_x is the general mean of the X-measures.

To test the hypothesis of equal adjusted means, it is necessary to obtain adjusted sums of squares and mean squares for the criterion variable. The adjusted total sum of squares may be calculated from:

$$SS'_{ty} = SS_{ty} - \frac{(SP_t)^2}{SS_{wx}}$$

where SS_{ty} is total sum of squares for Y.

Similarly, the adjusted sum of squares within groups may be determined from:

$$SS'_{wy} = SS_{wy} - \frac{(SP_w)^2}{SS_{wx}}$$

where SS_{wy} is sum of squares within for Y.

Finally, the adjusted sum of squares between groups may be calculated as a residual:

$$SS'_{by} = SS'_{ty} - SS'_{wy}$$

As in the simple analysis of variance, the number of degrees of freedom for the adjusted sum of squares between groups is $(k - 1)$. However, one degree of freedom is lost by imposing the restriction that the deviations be computed from the common within groups regression line, and the number of degrees of freedom for the adjusted sum of squares within groups is $(N - k - 1)$, and for the adjusted total sum of squares is $(N - 2)$. The adjusted mean squares between and within are, of course, obtained by dividing the sums of squares by their respective degrees of freedom. The test of the hypothesis of equal adjusted means is obtained from:

$$F = \frac{MS'_{by}}{MS'_{wy}} \quad \text{with } df = (k - 1), \quad (N - k - 1)$$

38.3 PROCEDURE

(1) Two or more random samples are independently drawn from the population under study, and these are subjected to different experimental treatments. The analysis of covariance may also be used to analyze existing differences between different populations.

(2) Prior to experimentation, a variable that is believed to be correlated with the criterion variable is selected for purposes of experimental control. A measure on this variable is recorded for each subject and later paired with the subject's measure on the criterion.

(3) At the conclusion of the experiment, a single criterion measure is administered to all subjects in all samples. The sums of squares between groups, within groups, and for total are calculated, along with their respective degrees of freedom. These calculations are completed for both the criterion and the control variable, and the calculations are precisely the same as those encountered earlier in the simple analysis of variance.

(4) The sums of products between groups, within groups, and for total are calculated in the manner outlined in Section 38.2; step-by-step procedures for calculating these quantities are given in Section 38.4. Adjusted sums of squares, degrees of freedom, and mean squares are calculated for the criterion, using the formulas given in Section 38.2. It is customary to record all of the sums of squares, sums of products, mean squares, and degrees of freedom in a summary table such as this:

Summary Table for the Analysis of Covariance							
Source	df	SS_x	SP	SS_y	df'	SS'_y	MS'_y
Between							
Within							
Total							XXX

(5) The F-ratio is calculated from:

$$F = \frac{MS'_{by}}{MS'_{wy}} \quad \text{with } df = (k - 1), \quad (N - k - 1)$$

The adjusted criterion means are calculated from:

$$\bar{Y}'_j = \bar{Y}_j - b_w(\bar{X}_j - M_x)$$

(6) The calculated F is compared to the tabled F at the desired level of significance and the appropriate number of degrees of freedom. If the calculated statistic equals or exceeds the tabled value, the null hypothesis is rejected, and a significant difference in the adjusted means of the various samples is determined to exist. If the calculated statistic is smaller than the tabled value, the null hypothesis is retained, and no significant difference in the adjusted means has been found.

38.4 COMPUTATIONAL PROCEDURES

It was noted earlier that the sums of squares and degrees of freedom for the X- and Y-measures are calculated in precisely the same fashion as encountered in the simple analysis of variance. These computations have been explained in great detail in a previous chapter (see Section 35.4).

The computations involved in the partitioning of the sum of products will be treated in detail in this section. The total sum of products has been identified as:

$$SP_t = \sum_{j=1}^{k} \sum_{i=1}^{n_j} X_{ij}Y_{ij} - \frac{T_x T_y}{N}$$

The formula for the sum of products between groups has been given as:

$$SP_b = \sum_{1=j}^{k} \frac{T_{xj}T_{yj}}{n_j} - \frac{T_x T_y}{N}$$

And the sum of products within groups may be calculated as a residual:

$$SP_w = SP_t - SP_b$$

Notice that only three terms need be evaluated to complete the calculations. Consider the first term of the total sum of products:

$$\sum_{j=1}^{k} \sum_{i=1}^{n_j} X_{ij}Y_{ij}$$

To evaluate this term, one need only (a) pair each X-score in the entire collection with its corresponding Y-score, and take the product of each pair and (b) sum all of these products.

Next, consider the second term of the total sum of products and of the sum of products between:

$$\frac{T_x T_y}{N}$$

To evaluate this term, (a) take the sum of all of the X-measures in the entire collection, and do the same for the Y-measures, (b) take the product of these two sums, and (c) divide by N.

Finally, to evaluate the first term of the sum of products between:

$$\sum_{j=1}^{k} \frac{T_{xj} T_{yj}}{n_j}$$

(a) take the sum of the X-measures for each group in the collection, and do the same for the Y-measures, (b) pair each X-sum with the Y-sum for the same group, and take the product for each group, (c) divide each product by the number of subjects in the group, and (d) sum for all groups.

The adjusted sums of squares, degrees of freedom, and mean squares may be calculated as outlined in Section 37.2.

38.5 EXAMPLE

Consider the data in the following table. Group membership constitutes the independent variable whose effects are to be studied, the X-score is the independent variable whose effects are to be controlled, and the Y-score is the dependent variable.

Group I						Group II				
X	Y	X^2	Y^2	XY		X	Y	X^2	Y^2	XY
0	3	0	9	0		2	2	4	4	4
1	5	1	25	5		3	4	9	16	12
3	5	9	25	15		5	4	25	16	20
4	6	16	36	24		6	5	36	25	30
4	8	16	64	32		6	7	36	49	42
6	9	36	81	54		8	8	64	64	64
18	36	78	240	130		30	30	174	174	172

(1) H: The two populations means are equal when the effects of the X-variable are controlled.

(2) Analysis of covariance, $n_1 = n_2 = 6$

(3) .05 level, $df = 1, 9$.
    R: $F \geqslant 5.12$.

(4) $SS_{bx} = 204 - 192 = 12$ $SS_{by} = 366 - 363 = 3$
     $SS_{wx} = 252 - 204 = 48$ $SS_{wy} = 414 - 366 = 48$
     $SS_{tx} = 252 - 192 = 60$ $SS_{ty} = 414 - 363 = 51$
     $SP_b = 258 - 264 = -6$ $SS'_{by} = 26.93 - 7.67 = 19.26$
     $SP_w = 302 - 258 = 44$ $SS'_{wy} = 48.00 - 40.33 = 7.67$
     $SP_t = 302 - 264 = 38$ $SS'_{ty} = 51.00 - 24.07 = 26.93$

Summary Table for the Analysis of Covariance							
Source	DF	SS_x	SP	SS_y	DF'	SS'_y	MS'_y
Between	1	12	−6	3	1	19.26	19.26
Within	10	48	44	48	9	7.67	.85
Total	11	60	38	51	10	26.93	XXX

$$F = \frac{19.26}{0.85} = 22.6, \, df = 1, 9$$

Means:

$$\bar{X}_1 = 3 \qquad \bar{Y}_1 = 6 \qquad \bar{Y}'_1 = 6 - .92(3 - 4) = 6.92$$
$$\bar{X}_2 = 5 \qquad \bar{Y}_2 = 5 \qquad \bar{Y}'_2 = 5 - .92(5 - 4) = 4.08$$

(5) Reject the null hypothesis. The adjusted mean for Group I is significantly larger than that of Group II. Notice that a simple analysis of variance performed on the unadjusted Y-measures would have yielded a nonsignificant finding ($F = .6$).

38.6 DISCUSSION

The analysis of covariance is one of the most valuable tools of statistical inference. It is especially appropriate in human learning research, where, to a large extent, the subjects' performance on the criterion may be determined before the experiment is undertaken. It is convenient to use, and it fits a wide variety of research situations.

The assumptions underlying the analysis of covariance are essentially the same as those for the analysis of variance, with the added assumption of homogeneity of regression. A test for homogeneity of regression may be developed by partitioning the adjusted sum of squares within into two parts:

$$SS_{wr} = SS_{wy} - \sum_{j=1}^{k} \frac{SP_j^2}{SS_{xj}} \quad \text{with } df = (N - 2k)$$

$$SS_{br} = SS'_{wy} - SS_{wr} \quad \text{with } df = (k - 1)$$

These sums of squares may be converted into mean squares by dividing by their respective degrees of freedom, and an F-ratio may be calculated from:

$$F = \frac{MS_{br}}{MS_{wr}} \quad \text{with } df = (k - 1), (N - 2k)$$

The test of the null hypothesis (no difference in the slopes of the regression lines) is accomplished by comparing the calculated F to the tabled value in the usual fashion.

The analysis of covariance may be extended to situations requiring the control of more than one variable or to the situation in which the regression is curvilinear through the use of multiple regression techniques, which are discussed in the following chapter.

EXERCISES

38.1 Consider the data in the following table:

	Control group		Experimental group	
SUBJECT	PRETEST	POSTTEST	PRETEST	POSTTEST
1	3	4	2	6
2	4	6	1	5
3	0	3	2	5
4	1	3	3	6
5	2	4		
Totals	10	20	8	22
Means	2	4	2	5.5

(a) Test the hypothesis of equal means for the control and experimental groups. Use the .05 level.

(b) Calculate the adjusted means for the two groups.

38.2 Refer to the data of Exercise 37.1. Convert sex to a binary variable by awarding a score of one to each male and a score of zero to each female. Complete an analysis of covariance summary table using urban versus rural group membership as the independent variable whose effects are to be studied, sex as the variable whose effects are to be controlled, and course grade as the dependent variable.

(a) Test the hypothesis of equal means for the urban and rural groups. Use the .05 level.

(b) Calculate the adjusted means for the two groups.

39

MULTIPLE REGRESSION

39.1 INTRODUCTION

Human behavior is an exceedingly complex affair, and the behavioral scientist will ordinarily expect to find a multitude of factors influencing any given action. Stated statistically, variation in a given dependent variable is usually a function of concomitant variation in many independent variables acting simultaneously. We have noted that academic achievement is in part determined by the variable commonly referred to as intelligence. Thus, the educational researcher should be able to predict student grades in a college composition class from intelligence test scores. He should also be able to predict these same grades from other variables such as rank in high school graduating class, grade in high school composition class, or score on a college entrance examination. Each of these variables is an attempt to assess a different facet of the student's total personality, and each could be expected to make some contribution toward future academic performance. The techniques of multiple regression enable the behavioral scientist to use his knowledge of two or more independent variables to predict scores on a single dependent variable with greater success than is possible with a knowledge of a single independent variable. In the example cited above, the investigator should be able to do a better job of predicting performance in the college class from a knowledge of performance in the corresponding high school class plus intelligence test score than from either of these variables taken alone.

Generally, the different kinds of variables encountered by the behavioral scientist will be correlated with each other, and to the extent that two variables are correlated, they may be said to measure the same thing.

264

Thus, there will be some overlap in the contribution of two-predictor variables in the typical multiple regression situation. Referring again to the example above, let us suppose that grade in the high school class accounts for 40 percent of the variance in college class grade, and that intelligence score accounts for 30 percent of the variance in the college grade. When high school grade and intelligence score are combined in a multiple regression equation, it is highly improbable that the investigator will be able to account for 70 percent of the variance in the college grade, but he can anticipate that he will be able to account for something between 40 and 70 percent. Because of this intercorrelation between the predictor variables, it is often expedient to use the best combination of two or three predictors, and there is little advantage to be gained from using additional predictor variables. For this reason, and because the computations become increasingly laborious with added variables, only two- and three-predictor regression equations will be treated in this chapter.

39.2 THE TWO-PREDICTOR CASE

The general form of the regression equation for raw scores in the two-predictor case is:

$$\hat{Y} = b_1 X_1 + b_2 X_2 + c$$

The same relationship may be expressed in deviation score form by:

$$\hat{y} = b_1 x_1 + b_2 x_2$$

As with the single-predictor case, the least squares criterion will be used to determine the appropriate values of b_1, b_2, and c. The derivation of the least squares solution requires the use of differential calculus; however, the problem has already been solved, and we shall be able to determine the prediction equation through the use of simple algebra. In the two-predictor situation using raw scores, there are three unknowns (b_1, b_2, and c) and three equations that must be solved simultaneously:

$$\Sigma X_1 Y = b_1 \Sigma X_1^2 + b_2 \Sigma X_1 X_2 + c \Sigma X_1$$
$$\Sigma X_2 Y = b_1 \Sigma X_1 X_2 + b_2 \Sigma X_2^2 + c \Sigma X_2$$
$$\Sigma Y = b_1 \Sigma X_1 + b_2 \Sigma X_2 + NC$$

These are sometimes referred to as *conditional* or *normal* equations. The problem may be simplified somewhat by using deviation scores. Then there are only two unknowns (b_1 and b_2), and two equations to be solved simultaneously:

$$\Sigma x_1 y = b_1 \Sigma x_1^2 + b_2 \Sigma x_1 x_2$$
$$\Sigma x_2 y = b_1 \Sigma x_1 x_2 + b_2 \Sigma x_2^2$$

The student is reminded that the values of the regression coefficients (b_1 and b_2) are the same whether deviation or raw scores are used. For this reason, some computational effort may be saved if the conditional equations are solved using deviation scores even though the end product is a regression equation using raw scores.

For the student who is unprepared to solve equations simultaneously, the author has derived the following formulas for b_1, b_2, and c:

$$b_1 = \frac{(\Sigma\, x_1 y)(\Sigma\, x_2^2) - (\Sigma\, x_1 x_2)(\Sigma\, x_2 y)}{(\Sigma\, x_1^2)(\Sigma\, x_2^2) - (\Sigma\, x_1 x_2)^2}$$

$$b_2 = \frac{(\Sigma\, x_1^2)(\Sigma\, x_2 y) - (\Sigma\, x_1 x_2)(\Sigma\, x_1 y)}{(\Sigma\, x_1^2)(\Sigma\, x_2^2) - (\Sigma\, x_1 x_2)^2}$$

$$c = \frac{\Sigma\, Y - b_1\, \Sigma\, X_1 - b_2\, \Sigma\, X_2}{N}$$

where:

$$\Sigma\, x_1^2 = \Sigma\, X_1{}^2 - \frac{(\Sigma\, X_1)^2}{N}$$

$$\Sigma\, x_2^2 = \Sigma\, X_2{}^2 - \frac{(\Sigma\, X_2)^2}{N}$$

$$\Sigma\, x_1 x_2 = \Sigma\, X_1 X_2 - \frac{(\Sigma\, X_1)(\Sigma\, X_2)}{N}$$

$$\Sigma\, x_1 y = \Sigma\, X_1 Y - \frac{(\Sigma\, X_1)(\Sigma\, Y)}{N}$$

$$\Sigma\, x_2 y = \Sigma\, X_2 Y - \frac{(\Sigma\, X_2)(\Sigma\, Y)}{N}$$

Consider the data of Table 39.1. The dependent variable (Y) and the first independent variable (X_2) are those used to illustrate the single-predictor situation (see Section 14.6). A second independent variable (X_2) in binary form has been added to illustrate the two-predictor situation. The least squares solution for the regression equation follows:

$$b_1 = \frac{(53)(2.5) - (10)(8)}{(90)(2.5) - (10)^2} = .42$$

$$b_2 = \frac{(90)(8) - (10)(53)}{(90)(2.5) - (10)^2} = 1.52$$

$$c = \frac{50 - .42(50) - 1.52(5)}{10} = 2.14$$

$$\hat{Y} = .42X_1 + 1.52X_2 + 2.14$$

Table 39.1 Data for the Two-Predictor Example

Y	X_1	X_2	Y^2	$X_1{}^2$	$X_2{}^2$	X_1Y	X_2Y	X_1X_2
2	1	0	4	1	0	2	0	0
3	2	0	9	4	0	6	0	0
2	2	0	4	4	0	4	0	0
5	3	1	25	9	1	15	5	3
5	4	0	25	16	0	20	0	0
7	5	1	49	25	1	35	7	5
5	6	0	25	36	0	30	0	0
6	8	1	36	64	1	48	6	8
7	9	1	49	81	1	63	7	9
8	10	1	64	100	1	80	8	10
50	50	5	290	340	5	303	33	35

39.3 THE THREE-PREDICTOR CASE

The general form of the regression equation for raw scores in the three-predictor case is:

$$\hat{Y} = b_1X_1 + b_2X_2 + b_3X_3 + c$$

The same relationship may be expressed in deviation score form by:

$$\hat{y} = b_1x_1 + b_2x_2 + b_3x_3$$

As with the single- and two-predictor cases, the least squares criterion will be used to determine the appropriate values of b_1, b_2, b_3, and c. The conditional equations are:

$$\Sigma\, X_1Y = b_1\, \Sigma\, X_1^2 + b_2\, \Sigma\, X_1X_2 + b_3\, \Sigma\, X_1X_3 + c\, \Sigma\, X_1$$
$$\Sigma\, X_2Y = b_1\, \Sigma\, X_1X_2 + b_2\, \Sigma\, X_2^2 + b_3\, \Sigma\, X_2X_3 + c\, \Sigma\, X_2$$
$$\Sigma\, X_3Y = b_1\, \Sigma\, X_1X_3 + b_2\, \Sigma\, X_2X_3 + b_3\, \Sigma\, X_3^2 + c\, \Sigma\, X_3$$
$$\Sigma\, Y = b_1\, \Sigma\, X_1 + b_2\, \Sigma\, X_2 + b_3\, \Sigma\, X_3 + Nc$$

These equations may be stated in deviation score form, eliminating one unknown (c) and one equation:

$$\Sigma\, x_1y = b_1\, \Sigma\, x_1^2 + b_2\, \Sigma\, x_1x_2 + b_3\, \Sigma\, x_1x_3$$
$$\Sigma\, x_2y = b_1\, \Sigma\, x_1x_2 + b_2\, \Sigma\, x_2^2 + b_3\, \Sigma\, x_2x_3$$
$$\Sigma\, x_3y = b_1\, \Sigma\, x_1x_3 + b_2\, \Sigma\, x_2x_3 + b_3\, \Sigma\, x_3^2$$

It is a relatively simple matter to extend the pattern that is evident in the two- and three-predictor cases to four or more predictors; how-

ever, the calculations become increasingly laborious and will probably
be undertaken only when an electronic computer is available.

39.4   NONLINEAR REGRESSION

Some of the relationships encountered by the behavioral scientist
are nonlinear, and the use of linear regression models in these situations
will grossly underestimate the degree of relationship and introduce
unnecessary prediction error. The most commonly encountered type
of nonlinear regression calls for a quadratic regression equation, an
equation that yields a curve with a single bend in it when plotted graphically. Such a relationship is graphically illustrated in Fig. 12.6.

The general form of the equation for quadratic regression is:

$$\hat{Y} = b_1 X + b_2 X^2 + c$$

If we represent X in this expression by X_1 and generate a dummy variable X_2 that is equal to X^2, the quadratic regression equation takes on
the form of the two-predictor linear regression equation:

$$\hat{Y} = b_1 X_1 + b_2 X_2 + c \qquad \text{where } X_2 = X_1^2$$

Treated in this fashion, the conditional equations for the quadratic
situation are the same as those for the two-predictor situation reported
earlier (see Section 39.2). We will use this approach to determine the
values of b_1, b_2, and c.

The data recorded in scattergram form in Fig. 12.3 suggest a slight
curvilinear relationship and will be used to illustrate quadratic regression.
These same data were used earlier to illustrate simple linear regression
(see Section 14.6).

The data of Fig. 12.3, along with the variable X_2, which has been
generated equal to X_1^2, have been recorded in Table 39.2. The least
squares solution for quadratic regression follows:

$$b_1 = \frac{(53)(11388) - (990)(553)}{(90)(11388) - (990)^2} = 1.25$$

$$b_2 = \frac{(90)(553) - (990)(53)}{(90)(11388) - (990)^2} = -.06$$

$$c = \frac{50 - 62.58 + 20.48}{10} = .79$$

$$Y = 1.25X - .06X^2 + .79$$

The graph of this equation, along with the original data from which it was generated, is reported in Fig. 39.1.

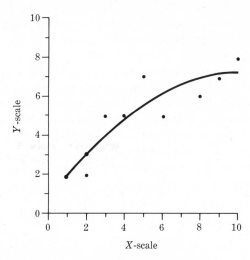

Figure 39.1   Curvilinear Regression.

Table 39.2   Data for the Quadratic Example

Y	X_1	X_2	Y^2	X_1^2	X_2^2	X_1Y	X_2Y	X_1X_2
2	1	1	4	1	1	2	2	1
3	2	4	9	4	16	6	12	8
2	2	4	4	4	16	4	8	8
5	3	9	25	9	81	15	45	27
5	4	16	25	16	256	20	80	64
7	5	25	49	25	625	35	175	125
5	6	36	25	36	1296	30	180	216
6	8	64	36	64	4096	48	384	512
7	9	81	49	81	6561	63	567	729
8	10	100	64	100	10000	80	800	1000
50	50	340	290	340	22948	303	2253	2690

39.5   THE MULTIPLE CORRELATION COEFFICIENT

The Pearson product moment correlation coefficient has proved extremely useful as an index of relationship between two variables, which we customarily designate X and Y, with X the independent variable and Y the dependent variable. In the case of simple linear regression, $\hat{Y}$ is a linear function of X; therefore, the Pearson correlation between $\hat{Y}$ and Y is precisely the same as that between X and Y. This suggests the possibility of developing a multiple correlation coefficient that is the correlation between $\hat{Y}$ and Y. This multiple correlation coefficient shall be represented by R.

The usefulness of the squared correlation coefficient, also referred to as the coefficient of determination, has already been noted. As in the simple linear prediction situation, the square of the multiple correlation coefficient (R^2) is the proportion of the variance of the dependent variable that is accounted for by the predictor variables. Thus, one may derive a general expression for the coefficient of determination that fits all regression models:

$$R^2 = \frac{\Sigma\,(\hat{Y} - \bar{Y})^2}{\Sigma\,(Y - \bar{Y})^2}$$

Because the $\hat{Y}$-values regress toward the mean, the value of R^2 will never exceed unity.

With a little effort, one may demonstrate that the square of the Pearson product moment correlation coefficient is calculable from the following expression:

$$r^2 = \frac{b\,\Sigma\,xy}{\Sigma\,y^2}$$

A similar expression may be derived that provides a convenient means of calculating the value of R^2 for the two-predictor and quadratic regression models:

$$R^2 = \frac{b_1\,\Sigma\,x_1 y + b_2\,\Sigma\,x_2 y}{\Sigma\,y^2}$$

The idea may be extended to the three-predictor model by:

$$R^2 = \frac{b_1\,\Sigma\,x_1 y + b_2\,\Sigma\,x_2 y + b_3\,\Sigma\,x_3 y}{\Sigma\,y^2}$$

For purposes of illustration, the values of the coefficient of determination will be calculated for the single-predictor, the two-predictor, and the quadratic regression examples used earlier. For the single-predictor case:

$$r^2 = \frac{(.59)(63)}{(40)} = .78$$

The investigator was able to account for 78 percent of the variance of the dependent variable from a knowledge of the single independent variable. For the two-predictor case:

$$R^2 = \frac{(.42)(53) + (1.52)(8)}{(40)} = .89$$

The additional information gained from the second independent variable enabled the investigator to account for an additional 11% of the variance of the dependent variable. Finally, for the quadratic model:

$$R^2 = \frac{(1.25)(53) - (.06)(553)}{(40)} = .83$$

Notice the small difference (only 5 percent) between the simple linear and the quadratic models. This suggests that the relationship between the two variables is almost linear, and this is apparent from an examination of Fig. 39.1.

When the coefficient of determination is known, it is a relatively simple matter to calculate the standard error of estimate:

$$SE_{est} = S_y \sqrt{1 - R^2} \qquad \text{where } S_y = \sqrt{\frac{SS_y}{df}}$$

The number of degrees of freedom for S_y is equal to $N - 2$ for the single-predictor model, $N - 3$ for the two-predictor and quadratic models, and $N - 4$ for the three-predictor model. For a discussion of the uses of the standard error of estimate and of the assumptions underlying its use, the reader is directed to Section 14.7.

EXERCISES

39.1 Given the following collection of scores, in which the X_1-variable is a score on an academic rating scale, X_2 is IQ score, and Y is grade received in a certain course of study.

Y	X_1	X_2
4	2	120
3	1	110
2	0	100
0	0	90
1	0	80

    (a)   Determine the regression equation for the two-predictor model, using X_1 and X_2 as independent variables.

    (b)   Calculate the coefficient of determination.

    (c)   Calculate the standard error of estimate.

39.2   (a)   Determine the regression equation for the single-predictor model, using X_1 as the independent variable.

    (b)   Calculate the coefficient of determination.

    (c)   Calculate the standard error of estimate.

    (d)   Determine the regression equation for the quadratic model, using X_1 as the independent variable.

    (e)   Calculate the coefficient of determination.

    (f)   Calculate the standard error of estimate.

    (g)   Plot the regression lines for both the linear and the quadratic regression equations.

THE ANALYSIS OF REGRESSION

40.1 INTRODUCTION

The analysis of regression provides a test of the hypothesis of no relationship between the independent and the dependent variables in a regression equation. In the single-predictor situation, the analysis of regression is equivalent to the test of the Pearson correlation coefficient cited in Section 30.2. In the multiple regression situation, the analysis of regression provides a test of the hypothesis of no relationship between the combined independent variables and the dependent variable—it may be thought of as a test of the hypothesis of no relationship between the Y-scores and the $\hat{Y}$-scores.

In the analysis of regression, the total sum of squares (SS_t) for the dependent variable is partitioned into a sum of squares for regression (SS_{reg}) and a sum of squares for error, or sum of squares for residuals (SS_{res}) as it is often called. This is equivalent to the partitioning of the total sum of squares in the simple analysis of variance; the sum of squares for regression corresponds to the sum of squares between groups, and the sum of squares for residuals corresponds to the sum of squares within groups. These are divided by their respective degrees of freedom to obtain the mean square for regression (MS_{reg}) and the mean square for residuals (MS_{res}); the F-ratio is formed with the residual mean square in the denominator.

The coefficient of determination (R^2) has been previously identified as that proportion of the variance of the dependent variable which is accounted for by the predictor variable or variables. Stated mathematically:

$$R^2 = \frac{SS_{reg}}{SS_t} \quad or \quad SS_{reg} = R^2 \, \Sigma \, y^2$$

Similarly, the coefficient of nondetermination $(1 - R^2)$ is that proportion of the variance of the dependent variable which is not accounted for by the predictor variables:

$$(1 - R^2) = 1 - \frac{SS_{reg}}{SS_t} = \frac{SS_{res}}{SS_t} \quad or \quad SS_{res} = (1 - R^2) \, \Sigma \, y^2$$

40.2 ANALYSIS OF REGRESSION WITH A SINGLE PREDICTOR

In the interest of simplicity, the formulas used in the analysis of regression are presented in deviation form. If the student has any doubts about the symbols used in these formulas, he should refer to Chap. 39. The formulas for the single-predictor case are:

$$SS_{reg} = \Sigma \, (\hat{Y} - \bar{Y})^2 = \Sigma \, \hat{y}^2 = b \, \Sigma \, xy \quad \text{with } df = 1$$
$$SS_{res} = \Sigma \, (Y - \hat{Y})^2 = \Sigma \, e^2 = SS_t - SS_{reg} \quad \text{with } df = N - 2$$
$$SS_t = \Sigma \, (Y - \bar{Y})^2 = \Sigma \, y^2 \quad \text{with } df = N - 1$$

$$MS_{reg} = \frac{SS_{reg}}{1}$$

$$MS_{res} = \frac{SS_{res}}{N - 2}$$

$$F = \frac{MS_{reg}}{MS_{res}} \quad \text{with } df = 1, (N - 2)$$

The single-predictor case is illustrated using the data of Section 14.6:

$$SS_{reg} = b \, \Sigma \, xy = 0.59(53) = 31.27$$
$$SS_{res} = SS_t - SS_{reg} = 40.00 - 31.27 = 8.73$$
$$MS_{reg} = 31.27$$

$$MS_{res} = \frac{8.73}{8} = 1.09$$

$$F = \frac{31.27}{1.09} = 28.7 \quad \text{with } df = 1, 8$$

The calculated F-ratio is significant at the 0.01 level, suggesting a relationship between the Y-scores and the $\hat{Y}$-scores. The investigator would conclude that the independent variable does in fact account for a portion of the variance of the dependent variable.

Occasionally, it will be convenient to calculate the F-ratio from this mathematically equivalent formula:

$$F = \frac{r^2}{\dfrac{(1 - r^2)}{(N - 2)}} = \frac{r^2(N - 2)}{(1 - r^2)}$$

A summary table for the analysis of regression may be constructed in the same fashion as for the analysis of variance. For the example:

Summary Table for the Analysis of Regression			
Sources	*DF*	*Sums of squares*	*Mean squares*
Regression	1	31.27	31.27
Residuals	8	8.73	1.09
Totals	9	40.00	XXXXX

40.3 ANALYSIS OF REGRESSION WITH MULTIPLE PREDICTORS

The formulas for the analysis of regression with two predictors are as follows:

$$SS_{reg} = \Sigma\, (\hat{Y} - \bar{Y})^2 = b_1 \Sigma\, x_1 y + b_2 \Sigma\, x_2 y \quad \text{with } df = 2$$

$$SS_{res} = \Sigma\, (Y - \hat{Y})^2 = SS_t - SS_{reg} \quad \text{with } df = N - 3$$

$$SS_t = \Sigma\, (Y - \bar{Y})^2 = \Sigma\, y^2 \quad \text{with } df = N - 1$$

$$MS_{reg} = \frac{SS_{reg}}{2}$$

$$MS_{res} = \frac{SS_{res}}{N - 3}$$

$$F = \frac{MS_{reg}}{MS_{res}} \quad \text{with } df = 2,\, (N - 3)$$

The two-predictor case is illustrated using the data of Section 39.2:

$$SS_{reg} = 0.42(53) + 0.52(8) = 35.42$$

$$SS_{res} = 40.00 - 35.42 = 4.58$$

$$MS_{reg} = \frac{35.42}{2} = 17.71$$

$$MS_{res} = \frac{4.58}{7} = 0.65$$

$$F = \frac{17.71}{0.65} = 27.2 \quad \text{with } df = 2, 7$$

The calculated F-ratio is significant at the 0.01 level, suggesting a relationship between the Y-scores and the $\hat{Y}$-scores. The investigator would conclude that the two independent variables combined in a least squares regression equation do in fact account for a portion of the variance of the dependent variable.

As before, the F-ratio may be calculated from a knowledge of the coefficient of determination:

$$F = \frac{\dfrac{R^2}{2}}{\dfrac{1 - R^2}{N - 3}}$$

The formulas developed for use with two predictors may also be used for the single-predictor quadratic case. In this situation X_2 is set equal to X^2 in the fashion set forth in Section 39.4.

The formulas for the analysis of regression with three predictors are as follows:

$$SS_{reg} = b_1 \Sigma x_1 y + b_2 \Sigma x_2 y + b_3 \Sigma x_3 y \quad \text{with } df = 3$$
$$SS_{res} = SS_t - SS_{reg} \quad \text{with } df = N - 4$$
$$SS_t = \Sigma y^2 \quad \text{with } df = N - 1$$

The mean squares and the F-ratio are formed as before. By now the pattern should be apparent, and the student should be able to extend the analysis of regression to any number of predictors.

40.4 COMPARING THE EFFECTIVENESS OF TWO PREDICTION EQUATIONS

The examples cited in the previous sections were drawn from the same data, the two-predictor case using the independent and the dependent variables from the single-predictor case plus one additional independent variable. The addition of a second predictor variable in this fashion cannot account for less of the variance than the single-predictor equation, but there may be a question as to whether it significantly increases the amount of variance accounted for. Stated differently, the addition of a second independent variable in the prediction equation cannot increase the error of prediction, but a test of significance may be desirable to determine whether the two-predictor equation produces significantly less error than the single-predictor equation. The test of significance to be outlined may be used to compare regression equations with any number of predictors.

The prediction equation using all of the predictors will be referred to as the full model (fm) and the equation with fewer predictors as the restricted model (rm). An adjusted sum of squares for regression will be calculated which is the difference between the sums of squares for regression of the full and the restricted models:

$$S_{\text{reg(diff)}} = SS_{\text{reg}(fm)} - SS_{\text{reg}(rm)}$$

The number of degrees of freedom for this difference sum of squares is the difference between the degrees of freedom for the full model and that for the restricted model. If the restricted model has one less predictor than the full model (as in our example), the difference sum of squares will have one degree of freedom. The error term for the test of significance of difference between the two models will be the mean square for residuals of the full model:

$$F = \frac{MS_{\text{reg(diff)}}}{MS_{\text{res}(fm)}}$$

Comparing the one- and two-predictor models used as examples in the previous sections:

$$SS_{\text{reg(diff)}} = 35.42 - 31.27 = 4.15 \quad \text{with } df = 1$$

$$F = \frac{4.15}{0.65} = 6.4 \quad \text{with } df = 1, 7$$

The calculated F-ratio is nonsignificant, indicating that the two-predictor equation is not significantly better than the single-predictor equation. In a practical setting, this would probably result in the adoption of the single-predictor equation.

If the coefficients of determination are known, the F-ratio may be calculated from:

$$F = \frac{\dfrac{R_{fm}^2 - R_{rm}^2}{u - v}}{\dfrac{1 - R_{fm}^2}{N - u - 1}}$$

where u is the number of predictors in the full model and v the number of predictors in the restricted model.

40.5 THE SIMPLE RANDOMIZED DESIGN

Consider the data from a very simple analysis of variance type problem:

Group	Scores	n_j	T_j	M_j	ΣX^2
I	0, 1, 2	3	3	1	5
II	0, 2, 4	3	6	2	20
III	2, 3, 4	3	9	3	29

$$\Sigma\Sigma X^2 = 5 + 20 + 29 = 54$$

$$\frac{T^2}{N} = \frac{18^2}{9} = 36$$

$$\frac{\Sigma T_j^2}{n_j} = \frac{9}{3} + \frac{36}{3} + \frac{81}{3} = 42$$

$$SS_b = 42 - 36 = 6$$
$$S_w = 54 - 42 = 12$$
$$SS_t = 54 - 36 = 18$$

Summary Table for the Analysis of Variance			
Sources	df	Sums of squares	Mean squares
Between	2	6	3
Within	6	12	2
Totals	8	18	XXX

$$F = \frac{3}{2} = 1.5 \text{ with } df = 2, 6$$

This type of problem may be translated into a least squares regression equation, and the analysis of regression yields the same result as the analysis of variance. Two predictor variables, X_1 and X_2, are generated to represent group membership (the independent variable in the analysis of variance). Members of Group I are awarded scores of one on X_1 and zero on X_2; members of Group II are awarded scores of zero on X_1 and one on X_2; members of Group III are awarded scores of zero on both X_1 and X_2. The data are recorded in Table 40.1 in a form convenient for completion of the analysis of regression.

Table 40.1   Data for the Simple Randomized Design

Y	X_1	X_2	Y^2	X_1^2	X_2^2	X_1Y	X_2Y	X_1X_2
0	1	0	0	1	0	0	0	0
1	1	0	1	1	0	1	0	0
2	1	0	4	1	0	2	0	0
0	0	1	0	0	1	0	0	0
2	0	1	4	0	1	0	2	0
4	0	1	16	0	1	0	4	0
2	0	0	4	0	0	0	0	0
3	0	0	9	0	0	0	0	0
4	0	0	16	0	0	0	0	0
18	3	3	54	3	3	3	6	0

The solution of the least squares regression equation is as follows:

$$\Sigma x_1^2 = 3 - \frac{9}{9} = 2$$

$$\Sigma x_2^2 = 3 - \frac{9}{9} = 2$$

$$\Sigma y^2 = 54 - \frac{324}{9} = 18$$

$$\Sigma x_1 y = 3 - \frac{54}{9} = -3$$

$$\Sigma x_2 y = 6 - \frac{54}{9} = 0$$

$$\Sigma x_1 x_2 = 0 - \frac{9}{9} = -1$$

$$b_1 = \frac{(-3)(2) - (-1)(0)}{(2)(2) - (-1)^2} = \frac{-6}{3} = -2$$

$$b_2 = \frac{(2)(0) - (-1)(-3)}{(2)(2) - (-1)^2} = \frac{-3}{3} = -1$$

$$c = \frac{18 - 3(-2) - 1(-3)}{9} = \frac{27}{9} = 3$$

$$\hat{Y} = -2X_1 - X_2 + 3$$

To predict the score of any individual, it is only necessary to know to which of the three groups he belongs:

For members of Group I: $\hat{Y} = -2 - 0 + 3 = 1$
For members of Group II: $\hat{Y} = -0 - 1 + 3 = 2$
For members of Group III: $\hat{Y} = -0 - 0 + 3 = 3$

Notice that the predicted score for any individual is the mean for his group on the dependent variable.

A summary table for the analysis of regression may be completed in the usual fashion:

$$SS_{reg} = (-2)(-3) + (-1)(0) = 6 \quad \text{with } df = 2$$
$$SS_{res} = 18 - 6 = 12 \quad \text{with } df = 6$$

Summary Table for the Analysis of Regression			
Sources	*df*	*Sums of squares*	*Mean squares*
Regression	2	6	3
Residuals	6	12	2
Totals	8	18	XXX

$$F = \frac{3}{2} = 1.5 \quad \text{with } df = 2, 6$$

The results, of course, are precisely the same as those achieved with the analysis of variance completed earlier. In other words, the analysis of regression completed in this fashion provides a test of the hypothesis of equal means. These procedures may be extended to multi-dimensional designs and to the analysis of covariance. When a computer is available, the use of regression analysis permits the investigator to develop complex designs using large numbers of variables. This is often desirable in research involving human subjects due to the complexity of human behavior and its determinants.

Most investigators using regression analysis as an alternative to the analysis of variance would have generated a third predictor, X_3, for this problem. Members of Groups I and II are awarded zeros on X_3, while members of Group III are awarded scores of one on X_3. Because X_3 contains no information not already included in X_1 and X_2, it does not alter the predicted scores or the analysis of regression, nor does it add another degree of freedom to the sum of squares for regression. The addition of this type of variable results in an infinity of solutions which meet the least squares criterion. The original regression equation suggests one of these:

$$\hat{Y} = -2X_1 - X_2 + 0X_3 + 3$$

Another interesting solution which meets the least squares criterion uses the group means as regression coefficients and eliminates the need for the constant:

$$\hat{Y} = X_1 + 2X_2 + 3X_3$$

Each of the three regression equations yields the group mean as the predicted score for each individual in the group.

Table 40.2   Detailed Analysis of the Predicted Scores

Y	y	y^2	$\hat{Y}$	$\hat{y}$	$\hat{y}^2$	e	e^2
0	−2	4	1	−1	1	−1	1
1	−1	1	1	−1	1	0	0
2	0	0	1	−1	1	+1	1
0	−2	4	2	0	0	−2	4
2	0	0	2	0	0	0	0
4	+2	4	2	0	0	+2	4
2	0	0	3	+1	1	−1	1
3	+1	1	3	+1	1	0	0
4	+2	4	3	+1	1	+1	1
18	0	18	18	0	6	0	12

Table 40.2 presents in detail an analysis of the predicted scores. The reader is reminded of the following relationships, which were used to complete the table:

$$y = Y - \bar{Y} \qquad \hat{y} = \hat{Y} - \bar{Y} \qquad e = Y - \hat{Y}$$

A number of interesting relationships are apparent from even a casual examination of the entries in the table. Special note should be taken of the column totals, which include the sums of squares for regression, residuals, and total.

40.6 THE ANALYSIS OF COVARIANCE

Regression analysis especially lends itself to the sort of problem encountered in Chap. 38—The Analysis of Covariance. The approach is surprising simple, even with many covariates: (1) One set of predictors is designated independent variables whose effects are to be studied: these will usually be binary variables organized in the fashion described in Section 40.5. (2) One set of predictors is designated independent variables whose effects are to be controlled; these may be either binary or continuous variables. (3) The full model includes all of the predictors. (4) The restricted model includes only those predictors whose effects are to be controlled. (5) An analysis of regression comparison of the two models yields the same results as the traditional analysis of covariance. When a computer is available, a wide variety of interesting hypotheses may be tested using this procedure.

The procedure is illustrated with the same problem used to illustrate the analysis of covariance in Chapter 38. The data have been arranged in Table 40.3 in a form convenient for the analysis of regression. The

Table 40.3   Data for the Analysis of Covariance

Y	X_1	X_2	Y^2	X_1^2	X_2^2	X_1Y	X_2Y	X_1X_2
3	0	1	9	0	1	0	3	0
5	1	1	25	1	1	5	5	1
5	3	1	25	9	1	15	5	3
6	4	1	36	16	1	24	6	4
8	4	1	64	16	1	32	8	4
9	6	1	81	36	1	54	9	6
2	2	0	4	4	0	4	0	0
4	3	0	16	9	0	12	0	0
4	5	0	16	25	0	20	0	0
5	6	0	25	36	0	30	0	0
7	6	0	49	36	0	42	0	0
8	8	0	64	64	0	64	0	0
66	48	6	414	252	6	302	36	18

variable X_1 is the original predictor variable (X in Chapter 38) whose effects are to be controlled, and X_2 is a binary variable representing group membership (1 for Group I, 0 for Group II).

The series of calculations which follows will be used in developing both the full and restricted models:

$$\Sigma\, y^2 = 414 - 363 = 51$$
$$\Sigma\, x_1^2 = 252 - 192 = 60$$
$$\Sigma\, x_2^2 = 6 - 3 = 3$$
$$\Sigma\, x_1 y = 302 - 264 = 38$$
$$\Sigma\, x_2 y = 36 - 33 = 3$$
$$\Sigma\, x_1 x_2 = 18 - 24 = -6$$

The full model is derived in the following fashion:

$$b_1 = \frac{(38)(3) - (-6)(3)}{(60)(3) - (-6)^2} = \frac{132}{144} = 0.917$$

$$b_2 = \frac{(60)(3) - (-6)(38)}{(60)(3) - (-6)^2} = \frac{408}{144} = 2.833$$

$$c = \frac{(66) - 0.917(48) - 2.833(6)}{12} = \frac{4.986}{12} = 0.416$$

$$\hat{Y} = 0.917 X_1 + 2.833 X_2 + 0.416$$
$$SS_{reg(fm)} = 0.917(38) + 2.833(3) = 43.345 \quad \text{with } df = 2$$
$$SS_{res(fm)} = 51.000 - 43.345 = 7.655 \quad \text{with } df = 9$$
$$R_{fm}^2 = \frac{43.345}{51} = 0.850$$

Now, the restricted model:

$$b = \frac{38}{60} = 0.633$$
$$c = 5.500 - 0.633(4) = 2.968$$
$$\hat{Y} = 0.633 X_1 + 2.968$$
$$SS_{reg(rm)} = 0.633(38) = 24.054 \quad \text{with } df = 1$$
$$R_{rm}^2 = \frac{24.054}{51} = 0.472$$

The two models may be compared using either the sums of squares-mean squares approach or the coefficients of determination approach outlined in Section 40.4. Using the sums of squares-mean squares:

$$F = \frac{\dfrac{43.345 - 24.054}{1}}{\dfrac{7.655}{9}} = \frac{19.29}{0.85} = 22.7 \qquad \text{with } df = 1, 9$$

Which is the same conclusion (within rounding error) which was arrived at using the traditional analysis of covariance.

APPENDIX

Table 1   A Table of Random Numbers*

Column number

Row	1	2	3	4	5	6	7	8	9	10	11	12	13	14	15	16	17	18	19	20	21	22	23	24	25	26	27	28	29	30	31	32
1	2	7	8	9	4	0	7	2	3	2	5	4	2	6	7	1	6	8	5	9	1	3	5	4	0	3	6	6	7	6	5	1
2	2	2	6	0	4	4	7	7	3	8	7	3	6	7	9	4	2	1	3	8	9	0	3	4	9	0	2	6	3	0	9	8
3	9	1	6	6	3	9	4	9	1	0	5	1	5	2	2	7	1	2	5	3	4	1	3	9	5	8	1	3	8	2	9	2
4	7	0	5	5	9	2	7	5	7	8	0	8	8	5	0	6	5	5	9	0	5	7	4	5	2	0	6	1	6	4	2	0
5	4	7	3	6	6	3	9	8	2	1	7	9	7	6	4	2	0	9	6	0	3	6	3	5	3	9	9	1	8	5	1	3
6	8	2	0	2	8	7	7	6	0	2	2	3	1	1	1	6	4	8	5	2	2	3	4	2	2	6	5	2	2	4	9	6
7	0	8	7	5	3	3	6	4	2	6	8	3	1	6	5	0	4	5	5	7	8	1	0	1	2	9	1	4	3	4	7	6
8	9	4	1	9	0	8	4	6	6	8	6	3	3	2	2	3	0	4	7	5	1	5	7	6	3	7	9	4	5	5	3	5
9	5	0	0	6	7	4	0	0	0	1	9	5	9	9	1	8	7	4	7	4	9	8	7	2	4	3	0	8	6	4	2	7
10	1	9	5	4	1	5	2	6	2	9	4	1	1	5	8	4	1	4	6	1	8	7	8	6	4	8	7	4	4	0	5	8
11	5	6	4	4	1	8	7	2	8	3	6	1	5	9	8	6	4	2	9	1	9	0	4	8	1	0	1	3	5	3	4	4
12	7	9	2	5	1	9	7	9	3	1	8	6	8	7	7	6	2	5	0	3	8	1	1	2	4	7	8	9	1	7	5	2
13	3	3	3	5	9	5	1	4	0	8	2	5	6	3	5	4	6	5	7	2	6	7	8	9	9	9	8	0	9	1	5	3
14	1	9	0	4	0	0	9	9	5	7	4	1	5	9	4	7	6	4	8	1	6	4	4	1	8	8	1	5	4	3	8	0
15	5	4	4	7	2	0	3	7	9	1	0	9	6	2	9	7	6	7	6	7	1	6	1	2	2	9	5	8	4	4	8	6
16	2	9	8	2	5	5	9	3	2	0	4	9	0	6	4	4	4	1	5	9	3	6	5	5	4	5	7	9	6	6	4	0
17	9	7	6	2	6	7	7	3	3	3	1	7	5	0	9	6	2	1	3	5	2	1	1	0	0	1	3	7	7	3	7	3
18	5	8	2	4	3	3	0	8	5	3	5	7	5	8	3	5	1	3	4	6	4	6	3	9	2	7	1	1	4	9	1	3
19	4	3	4	9	5	0	3	6	2	3	7	4	6	2	5	6	9	8	3	0	1	4	0	3	5	9	7	1	8	0	6	9
20	1	1	9	8	4	8	0	6	7	0	9	7	9	6	9	9	9	0	6	5	0	5	9	6	5	1	4	2	0	4	1	9
21	6	9	1	8	3	3	7	5	9	6	6	7	7	6	0	4	4	3	4	0	7	3	0	6	1	0	3	0	0	3	5	0
22	7	0	0	3	8	8	3	4	7	9	5	2	6	9	9	7	5	2	5	0	2	3	5	3	9	7	4	8	9	4	1	5
23	3	7	2	0	8	1	5	6	9	0	1	7	8	9	6	6	3	0	7	8	1	9	6	7	4	8	9	6	3	6	5	1

TABLE 1   287

24	25	26	27	28	29	30	31	32	33	34	35	36	37	38	39	40	41	42	43	44	45	46	47	48	49	50
2	2	8	3	2	9	6	9	0	1	2	1	0	9	8	4	2	6	9	3	4	7	1	6	8	7	0
1	1	6	0	7	0	6	5	4	0	3	7	6	9	3	9	1	0	3	6	1	8	0	8	5	7	2
5	3	2	7	2	1	3	7	4	9	0	8	8	8	3	5	3	4	5	3	0	4	1	7	7	1	5
5	6	1	0	0	3	8	5	5	6	8	8	6	3	4	6	7	0	8	9	9	2	0	7	8	6	4
4	4	7	9	2	1	6	1	1	1	6	7	2	4	8	0	0	5	7	1	1	9	6	7	2	1	6
2	8	6	5	7	3	2	7	4	8	4	2	6	4	4	2	7	4	0	7	9	7	4	3	3	9	3
4	9	1	6	7	5	5	1	5	2	5	9	2	7	4	1	6	5	1	2	0	4	1	0	6	0	3
4	0	7	6	8	8	9	6	0	2	0	8	2	3	8	0	1	3	3	8	1	9	9	1	3	0	3
0	7	1	4	2	8	4	5	4	9	9	5	7	8	3	2	2	8	5	7	4	0	5	8	1	1	5
9	7	4	3	0	3	3	5	1	5	6	3	7	6	3	3	4	7	8	3	6	0	6	3	5	1	9
4	1	4	3	2	9	1	5	8	6	0	0	4	2	7	0	1	8	9	1	4	6	7	3	7	6	8
8	8	4	1	9	8	2	1	7	8	9	3	2	6	4	1	1	6	1	5	6	4	3	8	9	1	5
2	3	5	6	5	0	7	2	3	3	1	1	3	2	0	4	8	1	2	8	0	3	1	5	5	2	2
2	4	7	7	2	7	5	1	9	1	6	3	8	1	8	7	7	5	9	6	8	6	0	6	3	2	4
7	0	4	9	5	0	8	6	3	6	2	0	5	0	4	0	0	0	2	8	6	1	9	4	2	0	9
1	6	2	8	5	1	8	6	1	4	5	6	0	3	2	3	4	4	6	7	5	1	2	9	0	3	0
3	9	2	1	9	3	4	5	3	9	4	1	5	7	0	5	6	8	2	1	4	7	5	7	6	5	9
9	4	4	2	0	7	4	3	5	1	3	1	2	0	8	7	0	7	8	7	9	2	2	7	1	9	9
2	7	6	7	1	8	0	8	9	9	9	4	4	3	7	1	2	4	4	7	8	8	6	5	4	7	3
3	4	4	4	3	2	0	1	0	0	6	0	6	2	1	8	8	3	0	0	1	0	2	2	8	2	6
0	2	3	6	2	9	8	7	5	7	8	5	7	6	9	1	9	8	5	4	4	5	4	0	5	5	1
6	8	2	9	6	3	2	9	5	7	5	7	6	1	7	2	6	8	1	5	7	6	2	0	8	1	7
5	2	2	0	3	2	8	3	4	3	8	5	1	8	6	7	9	9	8	5	2	5	8	1	8	0	9
6	4	5	7	7	0	9	9	1	4	2	9	6	8	2	0	4	3	2	2	3	8	3	2	2	2	5
0	9	7	5	8	9	6	6	5	1	3	8	7	1	6	8	9	8	2	8	6	0	3	1	7	5	0
5	4	8	3	0	0	5	6	9	4	6	8	1	8	9	8	9	2	9	2	5	5	5	5	0	8	7
6	8	1	4	7	2	9	2	9	8	0	8	8	8	3	6	8	7	0	5	3	7	2	4	2	0	5
0	7	3	1	1	2	9	9	7	1	1	1	8	2	1	9	0	1	6	4	7	6	4	0	3	4	4
0	0	3	2	0	6	6	2	7	8	5	0	7	3	8	4	7	0	3	3	9	6	9	4	0	3	6
0	7	9	9	7	6	5	5	7	5	9	2	9	9	2	6	8	0	0	8	6	8	0	8	6	8	3
7	0	6	2	4	3	9	5	4	7	3	9	2	5	2	7	8	3	4	2	4	5	6	8	3	4	8
2	3	6	5	0	9	2	8	8	7	8	8	0	2	2	0	4	9	0	1	0	0	3	0	4	3	0

Table 2   Squares, Square Roots, and Reciprocals

n	n^2	$\sqrt{n}$	$\sqrt{10n}$	$1/n$	n	n^2	$\sqrt{n}$	$\sqrt{10n}$	$1/n$
1	1	1.000	3.162	1.00000	51	2601	7.141	22.583	.01961
2	4	1.414	4.472	.50000	52	2704	7.211	22.804	.01923
3	9	1.732	5.477	.33333	53	2809	7.280	23.022	.01887
4	16	2.000	6.325	.25000	54	2916	7.348	23.238	.01852
5	25	2.236	7.071	.20000	55	3025	7.416	23.452	.01818
6	36	2.449	7.746	.16667	56	3136	7.483	23.664	.01786
7	49	2.646	8.367	.14286	57	3249	7.550	23.875	.01754
8	64	2.828	8.944	.12500	58	3364	7.616	24.083	.01724
9	81	3.000	9.487	.11111	59	3481	7.681	24.290	.01695
10	100	3.162	10.000	.10000	60	3600	7.746	24.495	.01667
11	121	3.317	10.488	.09091	61	3721	7.810	24.698	.01639
12	144	3.464	10.954	.08333	62	3844	7.874	24.900	.01613
13	169	3.606	11.402	.07692	63	3969	7.937	25.100	.01587
14	196	3.742	11.832	.07143	64	4096	8.000	25.298	.01562
15	225	3.873	12.247	.06667	65	4225	8.062	25.495	.01538
16	256	4.000	12.649	.06250	66	4356	8.124	25.690	.01515
17	289	4.123	13.038	.05882	67	4489	8.185	25.884	.01493
18	324	4.243	13.416	.05556	68	4624	8.246	26.077	.01471
19	361	4.359	13.784	.05263	69	4761	8.307	26.268	.01449
20	400	4.472	14.142	.05000	70	4900	8.367	26.458	.01429
21	441	4.583	14.491	.04762	71	5041	8.426	26.646	.01408
22	484	4.690	14.832	.04545	72	5184	8.485	26.833	.01389
23	529	4.796	15.166	.04348	73	5329	8.544	27.019	.01370
24	576	4.899	15.492	.04167	74	5476	8.602	27.203	.01351
25	625	5.000	15.811	.04000	75	5625	8.660	27.386	.01333
26	676	5.099	16.125	.03846	76	5776	8.718	27.568	.01316
27	729	5.196	16.432	.03704	77	5929	8.775	27.749	.01299
28	784	5.292	16.733	.03571	78	6084	8.832	27.928	.01282
29	841	5.385	17.029	.03448	79	6241	8.888	28.107	.01266
30	900	5.477	17.321	.03333	80	6400	8.944	28.284	.01250
31	961	5.568	17.607	.03226	81	6561	9.000	28.460	.01235
32	1024	5.657	17.889	.03125	82	6724	9.055	28.636	.01220
33	1089	5.745	18.166	.03030	83	6889	9.110	28.810	.01205
34	1156	5.831	18.439	.02941	84	7056	9.165	28.983	.01190
35	1225	5.916	18.708	.02857	85	7225	9.220	29.155	.01176
36	1296	6.000	18.974	.02778	86	7396	9.274	29.326	.01163
37	1369	6.083	19.235	.02703	87	7569	9.327	29.496	.01149
38	1444	6.164	19.494	.02632	88	7744	9.381	29.665	.01136
39	1521	6.245	19.748	.02564	89	7921	9.434	29.833	.01124
40	1600	6.325	20.000	.02500	90	8100	9.487	30.000	.01111
41	1681	6.403	20.248	.02439	91	8281	9.539	30.166	.01099
42	1764	6.481	20.494	.02381	92	8464	9.592	30.332	.01087
43	1849	6.557	20.736	.02326	93	8649	9.644	30.496	.01075
44	1936	6.633	20.976	.02273	94	8836	9.695	30.659	.01064
45	2025	6.708	21.213	.02222	95	9025	9.747	30.822	.01053
46	2116	6.782	21.448	.02174	96	9216	9.798	30.984	.01042
47	2209	6.856	21.679	.02128	97	9409	9.849	31.145	.01031
48	2304	6.928	21.909	.02083	98	9604	9.899	31.305	.01020
49	2401	7.000	22.136	.02041	99	9801	9.950	31.464	.01010
50	2500	7.071	22.361	.02000	100	10000	10.000	31.623	.01000

This table is reprinted from H. M. Walker and J. Lev, *Statistical Inference* (New York: Holt, Rinehart and Winston, Inc., 1953), p. 481.

TABLE 3   289

Table 3   Areas under the Normal Curve

Proportion of Total Area Under the Normal Curve Between Mean Ordinate and Ordinate at Given z Distance from the Mean

$\frac{x}{\sigma}$ or z	.00	.01	.02	.03	.04	.05	.06	.07	.08	.09
					Second decimal place in z					
.0	.0000	.0040	.0080	.0120	.0160	.0199	.0239	.0279	.0319	.0359
.1	.0398	.0438	.0478	.0517	.0557	.0596	.0636	.0675	.0714	.0753
.2	.0793	.0832	.0871	.0910	.0948	.0987	.1026	.1064	.1103	.1141
.3	.1179	.1217	.1255	.1293	.1331	.1368	.1406	.1443	.1480	.1517
.4	.1554	.1591	.1628	.1664	.1700	.1736	.1772	.1808	.1844	.1879
.5	.1915	.1950	.1985	.2019	.2054	.2088	.2123	.2157	.2190	.2224
.6	.2257	.2291	.2324	.2357	.2389	.2422	.2454	.2486	.2517	.2549
.7	.2580	.2611	.2642	.2673	.2704	.2734	.2764	.2794	.2823	.2852
.8	.2881	.2910	.2939	.2967	.2995	.3023	.3051	.3078	.3106	.3133
.9	.3159	.3186	.3212	.3238	.3264	.3289	.3315	.3340	.3365	.3389
1.0	.3413	.3438	.3461	.3485	.3508	.3531	.3554	.3577	.3599	.3621
1.1	.3643	.3665	.3686	.3708	.3729	.3749	.3770	.3790	.3810	.3830
1.2	.3849	.3869	.3888	.3907	.3925	.3944	.3962	.3980	.3997	.4015
1.3	.4032	.4049	.4066	.4082	.4099	.4115	.4131	.4147	.4162	.4177
1.4	.4192	.4207	.4222	.4236	.4251	.4265	.4279	.4292	.4306	.4319
1.5	.4332	.4345	.4357	.4370	.4382	.4394	.4406	.4418	.4429	.4441
1.6	.4452	.4463	.4474	.4484	.4495	.4505	.4515	.4525	.4535	.4545
1.7	.4554	.4564	.4573	.4582	.4591	.4599	.4608	.4616	.4625	.4633
1.8	.4641	.4649	.4656	.4664	.4671	.4678	.4686	.4693	.4699	.4706
1.9	.4713	.4719	.4726	.4732	.4738	.4744	.4750	.4756	.4761	.4767
2.0	.4772	.4778	.4783	.4788	.4793	.4798	.4803	.4808	.4812	.4817
2.1	.4821	.4826	.4830	.4834	.4838	.4842	.4846	.4850	.4854	.4857
2.2	.4861	.4864	.4868	.4871	.4875	.4878	.4881	.4884	.4887	.4890
2.3	.4893	.4896	.4898	.4901	.4904	.4906	.4909	.4911	.4913	.4916
2.4	.4918	.4920	.4922	.4925	.4927	.4929	.4931	.4932	.4934	.4936
2.5	.4938	.4940	.4941	.4943	.4945	.4946	.4948	.4949	.4951	.4952
2.6	.4953	.4955	.4956	.4957	.4959	.4960	.4961	.4962	.4963	.4964
2.7	.4965	.4966	.4967	.4968	.4969	.4970	.4971	.4972	.4973	.4974
2.8	.4974	.4975	.4976	.4977	.4977	.4978	.4979	.4979	.4980	.4981
2.9	.4981	.4982	.4982	.4983	.4984	.4984	.4985	.4985	.4986	.4986
3.0	.4987	.4987	.4987	.4988	.4988	.4989	.4989	.4989	.4990	.4990
3.1	.4990	.4991	.4991	.4991	.4992	.4992	.4992	.4992	.4993	.4993
3.2	.4993	.4993	.4994	.4994	.4994	.4994	.4994	.4995	.4995	.4995
3.3	.4995	.4995	.4995	.4996	.4996	.4996	.4996	.4996	.4996	.4997
3.4	.4997	.4997	.4997	.4997	.4997	.4997	.4997	.4997	.4997	.4998
3.5	.4998									
4.0	.49997									
4.5	.499997									
5.0	.4999997									

Example - 44.06 = z 1.56

Table 4   Ordinates of the Normal Curve

p	p	y	pq/y	p	p	y	pq/y
.01	.99	.027	.372	.26	.74	.324	.593
.02	.98	.048	.405	.27	.73	.331	.596
.03	.97	.068	.428	.28	.72	.337	.599
.04	.96	.086	.446	.29	.71	.342	.602
.05	.95	.103	.461	.30	.70	.348	.604
.06	.94	.119	.474	.31	.69	.353	.606
.07	.93	.134	.485	.32	.68	.358	.609
.08	.92	.149	.495	.33	.67	.362	.612
.09	.91	.162	.504	.34	.66	.366	.612
.10	.90	.176	.513	.35	.65	.370	.614
.11	.89	.188	.521	.36	.64	.374	.616
.12	.88	.200	.528	.37	.63	.378	.617
.13	.87	.212	.535	.38	.62	.381	.619
.14	.86	.223	.541	.39	.61	.384	.620
.15	.85	.233	.547	.40	.60	.386	.621
.16	.84	.243	.552	.41	.59	.389	.622
.17	.83	.253	.558	.42	.58	.391	.623
.18	.82	.262	.563	.43	.57	.393	.624
.19	.81	.271	.567	.44	.56	.394	.625
.20	.80	.280	.572	.45	.55	.396	.625
.21	.79	.288	.576	.46	.54	.397	.626
.22	.78	.296	.580	.47	.53	.398	.626
.23	.77	.304	.583	.48	.52	.398	.626
.24	.76	.311	.587	.49	.51	.399	.627
.25	.75	.318	.590	.50	.50	.399	.627

This table prepared by the author for this volume.

TABLE 5   291

Table 5 Tetrachoric Correlation Coefficients*

r_t	$\dfrac{bc}{ad}$	r_t	$\dfrac{bc}{ad}$	r_t	$\dfrac{bc}{ad}$
.00	0–1.00	.35	2.49–2.55	.70	8.50– 8.90
.01	1.01–1.03	.36	2.56–2.63	.71	8.91– 9.35
.02	1.04–1.06	.37	2.64–2.71	.72	9.36– 9.82
.03	1.07–1.08	.38	2.72–2.79	.73	9.83– 10.33
.04	1.09–1.11	.39	2.80–2.87	.74	10.34– 10.90
.05	1.12–1.14	.40	2.88–2.96	.75	10.91– 11.51
.06	1.15–1.17	.41	2.97–3.05	.76	11.52– 12.16
.07	1.18–1.20	.42	3.06–3.14	.77	12.17– 12.89
.08	1.21–1.23	.43	3.15–3.24	.78	12.90– 13.70
.09	1.24–1.27	.44	3.25–3.34	.79	13.71– 14.58
.10	1.28–1.30	.45	3.35–3.45	.80	14.59– 15.57
.11	1.31–1.33	.46	3.46–3.56	.81	15.58– 16.65
.12	1.34–1.37	.47	3.57–3.68	.82	16.66– 17.88
.13	1.38–1.40	.48	3.69–3.80	.83	17.89– 19.28
.14	1.41–1.44	.49	3.81–3.92	.84	19.29– 20.85
.15	1.45–1.48	.50	3.93–4.06	.85	20.86– 22.68
.16	1.49–1.52	.51	4.07–4.20	.86	22.69– 24.76
.17	1.53–1.56	.52	4.21–4.34	.87	24.77– 27.22
.18	1.57–1.60	.53	4.35–4.49	.88	27.23– 30.09
.19	1.61–1.64	.54	4.50–4.66	.89	30.10– 33.60
.20	1.65–1.69	.55	4.67–4.82	.90	33.61– 37.79
.21	1.70–1.73	.56	4.83–4.99	.91	37.80– 43.06
.22	1.74–1.78	.57	5.00–5.18	.92	43.07– 49.83
.23	1.79–1.83	.58	5.19–5.38	.93	49.84– 58.79
.24	1.84–1.88	.59	5.39–5.59	.94	58.80– 70.95
.25	1.89–1.93	.60	5.60–5.80	.95	70.96– 89.01
.26	1.94–1.98	.61	5.81–6.03	.96	89.02–117.54
.27	1.99–2.04	.62	6.04–6.28	.97	117.55–169.67
.28	2.05–2.10	.63	6.29–6.54	.98	169.68–293.12
.29	2.11–2.15	.64	6.55–6.81	.99	293.13–923.97
.30	2.16–2.22	.65	6.82–7.10	1.00	938.98–
.31	2.23–2.28	.66	7.11–7.42		
.32	2.29–2.34	.67	7.43–7.75		
.33	2.35–2.41	.68	7.76–8.11		
.34	2.42–2.48	.69	8.12–8.49		

* This table is reprinted from M. D. Davidoff and H. W. Goheen, "A table for the rapid determination of the tetrachoric correlation coefficient," *Psychometrika*, 1953, **18**, 115–121, by permission of the authors and the editors of *Psychometrika*.

To use the table, set the data up in a 2 × 2 table as shown in the text, Chap. 13. Enter the table with bc/ad or its reciprocal, whichever is the larger, and read the corresponding value of r_t. The accuracy of the values given for r_t does not extend beyond the second decimal, and interpolation between the values listed for bc/ad is not recommended.

Table 6 Coefficients of the Binomial Distribution

N	$\binom{N}{0}$	$\binom{N}{1}$	$\binom{N}{2}$	$\binom{N}{3}$	$\binom{N}{4}$	$\binom{N}{5}$	$\binom{N}{6}$	$\binom{N}{7}$	$\binom{N}{8}$	$\binom{N}{9}$	$\binom{N}{10}$
0	1										
1	1	1									
2	1	2	1								
3	1	3	3	1							
4	1	4	6	4	1						
5	1	5	10	10	5	1					
6	1	6	15	20	15	6	1				
7	1	7	21	35	35	21	7	1			
8	1	8	28	56	70	56	28	8	1		
9	1	9	36	84	126	126	84	36	9	1	
10	1	10	45	120	210	252	210	120	45	10	1
11	1	11	55	165	330	462	462	330	165	55	11
12	1	12	66	220	495	792	924	792	495	220	66
13	1	13	78	286	715	1287	1716	1716	1287	715	286
14	1	14	91	364	1001	2002	3003	3432	3003	2002	1001
15	1	15	105	455	1365	3003	5005	6435	6435	5005	3003
16	1	16	120	560	1820	4368	8008	11440	12870	11440	8008
17	1	17	136	680	2380	6188	12376	19448	24310	24310	19448
18	1	18	153	816	3060	8568	18564	31824	43758	48620	43758
19	1	19	171	969	3876	11628	27132	50388	75582	92378	92378
20	1	20	190	1140	4845	15504	38760	77520	125970	167960	184756

This table is reprinted from Sydney Siegel, *Nonparametric Statistics for the Behavioral Sciences*, (New York: McGraw-Hill Book Company, 1956), p. 288.

TABLE 7   293

Table 7   Distribution of *t* for Given Probability Levels

df	Level of significance for one-tailed test					
	.10	.05	.025	.01	.005	.0005
	Level of significance for two-tailed test					
	.20	.10	.05	.02	.01	.001
1	3.078	6.314	12.706	31.821	63.657	636.619
2	1.886	2.920	4.303	6.965	9.925	31.598
3	1.638	2.353	3.182	4.541	5.841	12.941
4	1.533	2.132	2.776	3.747	4.604	8.610
5	1.476	2.015	2.571	3.365	4.032	6.859
6	1.440	1.943	2.447	3.143	3.707	5.959
7	1.415	1.895	2.365	2.998	3.499	5.405
8	1.397	1.860	2.306	2.896	3.355	5.041
9	1.383	1.833	2.262	2.821	3.250	4.781
10	1.372	1.812	2.228	2.764	3.169	4.587
11	1.363	1.796	2.201	2.718	3.106	4.437
12	1.356	1.782	2.179	2.681	3.055	4.318
13	1.350	1.771	2.160	2.650	3.012	4.221
14	1.345	1.761	2.145	2.624	2.977	4.140
15	1.341	1.753	2.131	2.602	2.947	4.073
16	1.337	1.746	2.120	2.583	2.921	4.015
17	1.333	1.740	2.110	2.567	2.898	3.965
18	1.330	1.734	2.101	2.552	2.878	3.992
19	1.328	1.729	2.093	2.539	2.861	3.883
20	1.325	1.725	2.086	2.528	2.845	3.850
21	1.323	1.721	2.080	2.518	2.831	3.819
22	1.321	1.717	2.074	2.508	2.819	3.792
23	1.319	1.714	2.069	2.500	2.807	3.767
24	1.318	1.711	2.064	2.492	2.797	3.745
25	1.316	1.708	2.060	2.485	2.787	3.725
26	1.315	1.706	2.056	2.479	2.779	3.707
27	1.314	1.703	2.052	2.473	2.771	3.690
28	1.313	1.701	2.048	2.467	2.763	3.674
29	1.311	1.699	2.045	2.462	2.756	3.659
30	1.310	1.697	2.042	2.457	2.750	3.646
40	1.303	1.684	2.021	2.423	2.704	3.551
60	1.296	1.671	2.000	2.390	2.660	3.460
120	1.289	1.658	1.980	2.358	2.617	3.373
∞	1.282	1.645	1.960	2.326	2.576	3.291

* This table is abridged from Table III of R. A. Fisher and F. Yates: *Statistical Tables for Biological, Agricultural, and Medical Research,* published by Oliver and Boyd, Ltd., Edinburgh, by permission of the authors and publishers.

Table 8   Mann-Whitney U
One-tailed test at .05 level; two-tailed test at .10 level

m	\multicolumn{20}{c}{n}
	1	2	3	4	5	6	7	8	9	10	11	12	13	14	15	16	17	18	19	20
1	—																			
2	—	—																		
3	—	—	0																	
4	—	—	0	1																
5	—	0	1	2	4															
6	—	0	2	3	5	7														
7	—	0	2	4	6	8	11													
8	—	1	3	5	8	10	13	15												
9	—	1	4	6	9	12	15	18	21											
10	—	1	4	7	11	14	17	20	24	27										
11	—	1	5	8	12	16	19	23	27	31	34									
12	—	2	5	9	13	17	21	26	30	34	38	42								
13	—	2	6	10	15	19	24	28	33	37	42	47	51							
14	—	3	7	11	16	21	26	31	36	41	46	51	56	61						
15	—	3	7	12	18	23	28	33	39	44	50	55	61	66	72					
16	—	3	8	14	19	25	30	36	42	48	54	60	65	71	77	83				
17	—	3	9	15	20	26	33	39	45	51	57	64	70	77	83	89	96			
18	—	4	9	16	22	28	35	41	48	55	61	68	75	82	88	95	102	109		
19	0	4	10	17	23	30	37	44	51	58	65	72	80	87	94	101	109	116	123	
20	0	4	11	18	25	32	39	47	54	62	69	77	84	92	100	107	115	123	130	138
21	0	5	11	19	26	34	41	49	57	65	73	81	89	97	105	113	121	130	138	146
22	0	5	12	20	28	36	44	52	60	68	77	85	94	102	111	119	128	136	145	154
23	0	5	13	21	29	37	46	54	63	72	81	90	98	107	116	125	134	143	152	161
24	0	6	13	22	30	39	48	57	66	75	85	94	103	113	122	131	141	150	160	162
25	0	6	14	23	32	41	50	60	69	79	89	98	108	118	128	137	147	157	167	177
26	0	6	15	24	33	43	53	62	72	82	92	103	113	123	133	143	154	164	174	185
27	0	7	15	25	35	45	55	65	75	86	96	107	117	128	139	149	160	171	182	192
28	0	7	16	26	36	46	57	68	78	89	100	111	122	133	144	156	167	178	189	200
29	0	7	17	27	38	48	59	70	82	93	104	116	127	138	150	162	173	185	196	208
30	0	7	17	28	39	50	61	73	85	96	108	120	132	144	156	168	180	192	204	216
31	0	8	18	29	40	52	64	76	88	100	112	124	136	149	161	174	186	199	211	224
32	0	8	19	30	42	54	66	78	91	103	116	128	141	154	167	180	193	206	218	231
33	0	8	19	31	43	56	68	81	94	107	120	133	146	159	172	186	199	212	226	239
34	0	9	20	32	45	57	70	84	97	110	124	137	151	164	178	192	206	219	233	247
35	0	9	21	33	46	59	73	86	100	114	128	141	156	170	184	198	212	226	241	255
36	0	9	21	34	48	61	75	89	103	117	131	146	160	175	189	204	219	233	248	263
37	0	10	22	35	49	63	77	91	106	121	135	150	165	180	195	210	225	240	255	271
38	0	10	23	36	50	65	79	94	109	124	139	154	170	185	201	216	232	247	263	278
39	1	10	23	38	52	67	82	97	112	128	143	159	175	190	206	222	238	254	270	286
40	1	11	24	39	53	68	84	99	115	131	147	163	179	196	212	228	245	261	278	294

This table is reprinted from *American Statistical Association Journal* (September 1964), pp. 927–932.

TABLE 8   295

Table 8   Mann-Whitney U (*Continued*)
One-tailed test at .05 level; two-tailed test at .10 level

m										n										
	1	2	3	4	5	6	7	8	9	10	11	12	13	14	15	16	17	18	19	20
1	—																			
2	—	—																		
3	—	—	—																	
4	—	—	—	0																
5	—	—	0	1	2															
6	—	—	1	2	3	5														
7	—	—	1	3	5	6	8													
8	—	0	2	4	6	8	10	13												
9	—	0	2	4	7	10	12	15	17											
10	—	0	3	5	8	11	14	17	20	23										
11	—	0	3	6	9	13	16	19	23	26	30									
12	—	1	4	7	11	14	18	22	26	29	33	37								
13	—	1	4	8	12	16	20	24	28	33	37	41	45							
14	—	1	5	9	13	17	22	26	31	36	40	45	50	55						
15	—	1	5	10	14	19	24	29	34	39	44	49	54	59	64					
16	—	1	6	11	15	21	26	31	37	42	47	53	59	64	70	75				
17	—	2	6	11	17	22	28	34	39	45	51	57	63	69	75	81	87			
18	—	2	7	12	18	24	30	36	42	48	55	61	67	74	80	86	93	99		
19	—	2	7	13	19	25	32	38	45	52	58	65	72	78	85	92	99	106	113	
20	—	2	8	14	20	27	34	41	48	55	62	69	76	83	90	98	105	112	119	127
21	—	3	8	15	22	29	36	43	50	58	65	73	80	88	96	103	111	119	126	134
22	—	3	9	16	23	30	38	45	53	61	69	77	85	93	101	109	117	125	133	141
23	—	3	9	17	24	32	40	48	56	64	73	81	89	98	106	115	123	132	140	149
24	—	3	10	17	25	33	42	50	59	67	76	85	94	102	111	120	129	138	147	156
25	—	3	10	18	27	35	44	53	62	71	80	89	98	107	117	126	135	145	154	163
26	—	4	11	19	28	37	46	55	64	74	83	93	102	112	122	132	141	151	161	171
27	—	4	11	20	29	38	48	57	67	77	87	97	107	117	127	137	147	158	168	178
28	—	4	12	21	30	40	50	60	70	80	90	101	111	122	132	143	154	160	171	186
29	—	4	13	22	32	42	52	62	73	83	94	105	116	127	138	149	160	171	182	193
30	—	5	13	23	33	43	54	65	76	87	98	109	120	131	143	154	166	177	189	200
31	—	5	14	24	34	45	56	67	78	90	101	113	125	136	148	160	172	184	196	208
32	—	5	14	24	35	46	58	69	81	93	105	117	129	141	153	166	178	190	203	215
33	—	5	15	25	37	48	60	72	84	96	108	121	133	146	159	171	184	197	210	222
34	—	5	15	26	38	50	62	74	87	99	112	125	138	151	164	177	190	203	217	230
35	—	6	16	27	39	51	64	77	89	103	116	129	142	156	169	183	196	210	224	237
36	—	6	16	28	40	53	66	79	92	106	119	133	147	161	174	188	202	216	231	245
37	—	6	17	29	41	55	68	81	95	109	123	137	151	165	180	194	209	223	238	252
38	—	6	17	30	43	56	70	84	98	112	127	141	156	170	185	200	215	230	245	259
39	0	7	18	31	44	58	72	86	101	115	130	145	160	175	190	206	321	236	252	267
40	0	7	18	31	45	59	74	89	103	119	134	149	165	180	196	211	227	243	258	274

Table 8 Mann-Whitney U (*Continued*)
One-tailed test at .05 level; two-tailed test at .10 level

m	n=1	2	3	4	5	6	7	8	9	10	11	12	13	14	15	16	17	18	19	20
1	—																			
2	—	—																		
3	—	—	—																	
4	—	—	—	—																
5	—	—	—	0	1															
6	—	—	—	1	2	3														
7	—	—	0	1	3	4	6													
8	—	—	0	2	4	6	7	9												
9	—	—	1	3	5	7	9	11	14											
10	—	—	1	3	6	8	11	13	16	19										
11	—	—	1	4	7	9	12	15	18	22	25									
12	—	—	2	5	8	11	14	17	21	24	28	31								
13	—	0	2	5	9	12	16	20	23	27	31	35	39							
14	—	0	2	6	10	13	17	22	26	30	34	38	43	47						
15	—	0	3	7	11	15	19	24	28	33	37	42	47	51	56					
16	—	0	3	7	12	16	21	26	31	36	41	46	51	56	61	66				
17	—	0	4	8	13	18	23	28	33	38	44	49	55	60	66	71	77			
18	—	0	4	9	14	19	24	30	36	41	47	53	59	65	70	76	82	88		
19	—	1	4	9	15	20	26	32	38	44	50	56	63	69	75	82	88	94	101	
20	—	1	5	10	16	22	28	34	40	47	53	60	67	73	80	87	93	100	107	114
21	—	1	5	11	17	23	30	36	43	50	57	64	71	78	85	92	99	106	113	121
22	—	1	6	11	18	24	31	38	45	53	60	67	75	82	90	97	105	112	120	127
23	—	1	6	12	19	26	33	40	48	55	63	71	79	87	94	102	110	118	126	134
24	—	1	6	13	20	27	35	42	50	58	66	75	83	91	99	108	116	124	133	141
25	—	1	7	13	21	29	36	45	53	61	70	78	87	95	104	113	122	130	139	148
26	—	1	7	14	22	30	38	47	55	64	73	82	91	100	109	118	127	136	146	155
27	—	2	7	15	23	31	40	49	58	67	76	85	95	104	114	123	133	142	152	162
28	—	2	8	16	24	33	42	51	60	70	79	89	99	109	119	129	139	149	159	169
29	—	2	8	16	25	34	43	53	63	73	83	93	103	113	123	134	144	155	165	176
30	—	2	9	17	26	35	45	55	65	76	86	96	107	118	128	139	150	161	172	182
31	—	2	9	18	27	37	47	57	68	78	89	100	111	122	133	144	156	167	178	189
32	—	2	9	18	28	38	49	59	70	81	92	104	115	127	138	150	161	173	185	196
33	—	2	10	19	29	40	50	61	73	84	96	107	119	131	143	155	167	179	191	203
34	—	3	10	20	30	41	52	64	75	87	99	111	123	135	148	160	173	185	198	210
35	—	3	11	20	31	42	54	66	78	90	102	115	127	140	153	165	178	191	204	217
36	—	3	11	21	32	44	56	68	80	93	106	118	131	144	158	171	184	197	211	224
37	—	3	11	22	33	45	57	70	83	96	109	122	135	149	162	176	190	203	217	231
38	—	3	12	22	34	46	59	72	85	99	112	126	139	153	167	181	195	209	224.	238
39	—	3	12	23	35	48	61	74	88	101	115	129	144	158	172	187	201	216	230	245
40	—	3	13	24	36	49	63	76	90	104	119	133	148	162	177	192	207	222	237	252

TABLE 8   297

Table 8   Mann-Whitney U (*Continued*)
One-tailed test at .05 level; two-tailed test at .10 level

m	n																			
	1	2	3	4	5	6	7	8	9	10	11	12	13	14	15	16	17	18	19	20
1	—																			
2	—	—																		
3	—	—	—																	
4	—	—	—	—																
5	—	—	—	—	0															
6	—	—	—	0	1	2														
7	—	—	—	0	1	3	4													
8	—	—	—	1	2	4	6	7												
9	—	—	0	1	3	5	7	9	11											
10	—	—	0	2	4	6	9	11	13	16										
11	—	—	0	2	5	7	10	13	16	18	21									
12	—	—	1	3	6	9	12	15	18	21	24	27								
13	—	—	1	3	7	10	13	17	20	24	27	31	34							
14	—	—	1	4	7	11	15	18	22	26	30	34	38	42						
15	—	—	2	5	8	12	16	20	24	29	33	37	42	46	51					
16	—	—	2	5	9	13	18	22	27	31	36	41	45	50	55	60				
17	—	—	2	6	10	15	19	24	29	34	39	44	49	54	60	65	70			
18	—	—	2	6	11	16	21	26	31	37	42	47	53	58	64	70	75	81		
19	—	0	3	7	12	17	22	28	33	39	45	51	57	63	69	74	81	87	93	
20	—	0	3	8	13	18	24	30	36	42	48	54	60	67	73	79	86	92	99	105
21	—	0	3	8	14	19	25	32	38	44	51	58	64	71	78	84	91	98	105	112
22	—	0	4	9	14	21	27	34	40	47	54	61	68	75	82	89	96	104	111	118
23	—	0	4	9	15	22	29	35	43	50	57	64	72	79	87	94	102	109	117	125
24	—	0	4	10	16	23	30	37	45	52	60	68	75	83	91	99	107	115	123	131
25	—	0	5	10	17	24	32	39	47	55	63	71	79	87	96	104	112	121	129	138
26	—	0	5	11	18	25	33	41	49	58	66	74	83	92	100	109	118	127	135	144
27	—	1	5	12	19	27	35	43	52	60	69	78	87	96	105	114	123	132	142	151
28	—	1	5	12	20	28	36	45	54	63	72	81	91	100	109	119	128	138	148	157
29	—	1	6	13	21	29	38	47	56	66	75	85	94	104	114	124	134	144	154	164
30	—	1	6	13	22	30	40	49	58	68	78	88	98	108	119	129	139	150	160	170
31	—	1	6	14	22	32	41	51	61	71	81	92	102	113	123	134	145	155	166	177
32	—	1	7	14	23	33	43	53	63	74	84	95	106	117	128	139	150	161	172	184
33	—	1	7	15	24	34	44	55	65	76	87	98	110	121	132	144	155	167	179	190
34	—	1	7	16	25	35	46	57	68	79	90	102	113	125	137	149	161	173	185	197
35	—	1	8	16	26	37	47	59	70	82	93	105	117	129	142	154	166	179	191	203
36	—	1	8	17	27	38	49	60	72	84	96	109	121	134	146	159	172	184	197	210
37	—	1	8	17	28	39	51	62	75	87	99	112	125	138	151	164	177	190	203	217
38	—	1	9	18	29	40	52	64	77	90	102	116	129	142	155	169	182	196	210	223
39	—	2	9	19	30	41	54	66	79	92	106	119	133	146	160	174	188	202	216	230
40	—	2	9	19	31	43	55	68	81	95	109	122	136	150	165	179	193	208	222	237

Table 9   Wilcoxon Matched-pairs Signed-ranks Test*

N	Level of significance for one-tailed test		
	.025	.01	.005
	Level of significance for two-tailed test		
	.05	.02	.01
6	0	—	—
7	2	0	—
8	4	2	0
9	6	3	2
10	8	5	3
11	11	7	5
12	14	10	7
13	17	13	10
14	21	16	13
15	25	20	16
16	30	24	20
17	35	28	23
18	40	33	28
19	46	38	32
20	52	43	38
21	59	49	43
22	66	56	49
23	73	62	55
24	81	69	61
25	89	77	68

* Adapted from Table I of F. Wilcoxon, *Some Rapid Approximate Statistical Procedures*, p. 13, American Cyanamid Company, New York, 1949, with their kind permission.

TABLE 10   299

Table 10   Distribution of Chi-square for Given Probability Levels*

Degrees of freedom	Probability												
	.995	.990	.975	.950	.900	.750	.500	.250	.100	.050	.025	.010	.005
1	.0000393	.000157	.000982	.00393	.0158	.102	.455	1.32	2.71	3.84	5.02	6.63	7.88
2	.0100	.0201	.0506	.103	.211	.575	1.39	2.77	4.61	5.99	7.38	9.21	10.6
3	.0717	.115	.216	.352	.584	1.21	2.37	4.11	6.25	7.81	9.35	11.3	12.8
4	.207	.297	.484	.711	1.06	1.92	3.36	5.39	7.78	9.49	11.1	13.3	14.9
5	.412	.554	.831	1.15	1.61	2.67	4.35	6.63	9.24	11.1	12.8	15.1	16.7
6	.676	.872	1.24	1.64	2.20	3.45	5.35	7.84	10.6	12.6	14.4	16.8	18.5
7	.989	1.24	1.69	2.17	2.83	4.25	6.35	9.04	12.0	14.1	16.0	18.5	20.3
8	1.34	1.65	2.18	2.73	3.49	5.07	7.34	10.2	13.4	15.5	17.5	20.1	22.0
9	1.73	2.09	2.70	3.33	4.17	5.90	8.34	11.4	14.7	16.9	19.0	21.7	23.6
10	2.16	2.56	3.25	3.94	4.87	6.74	9.34	12.5	16.0	18.3	20.5	23.2	25.2
11	2.60	3.05	3.82	4.57	5.58	7.58	10.3	13.7	17.3	19.7	21.9	24.7	26.8
12	3.07	3.57	4.40	5.23	6.30	8.44	11.3	14.8	18.5	21.0	23.3	26.2	28.3
13	3.57	4.11	5.01	5.89	7.04	9.30	12.3	16.0	19.8	22.4	24.7	27.7	29.8
14	4.07	4.66	5.63	6.57	7.79	10.2	13.3	17.1	21.1	23.7	26.1	29.1	31.3
15	4.60	5.23	6.26	7.26	8.55	11.0	14.3	18.2	22.3	25.0	27.5	30.6	32.8
16	5.14	5.81	6.91	7.96	9.31	11.9	15.3	19.4	23.5	26.3	28.8	32.0	34.3
17	5.70	6.41	7.56	8.67	10.1	12.8	16.3	20.5	24.8	27.6	30.2	33.4	35.7
18	6.26	7.01	8.23	9.39	10.9	13.7	17.3	21.6	26.0	28.9	31.5	34.8	37.2
19	6.84	7.63	8.91	10.1	11.7	14.6	18.3	22.7	27.2	30.1	32.9	36.2	38.6
20	7.43	8.26	9.59	10.9	12.4	15.5	19.3	23.8	28.4	31.4	34.2	37.6	40.0

(continued on next page)

Table 10　Distribution of Chi-square for Given Probability Levels* (*Continued*)

Degrees of freedom	Probability												
	.995	.990	.975	.950	.900	.750	.500	.250	.100	.050	.025	.010	.005
21	8.03	8.90	10.3	11.6	13.2	16.3	20.3	24.9	29.6	32.7	35.5	38.9	41.4
22	8.64	9.54	11.0	12.3	14.0	17.2	21.3	26.0	30.8	33.9	36.8	40.3	42.8
23	9.26	10.2	11.7	13.1	14.8	18.1	22.3	27.1	32.0	35.2	38.1	41.6	44.2
24	9.89	10.9	12.4	13.8	15.7	19.0	23.3	28.2	33.2	36.4	39.4	43.0	45.6
25	10.5	11.5	13.1	14.6	16.5	19.9	24.3	29.3	34.4	37.7	40.6	44.3	46.5
26	11.2	12.2	13.8	15.4	17.3	20.8	25.3	30.4	35.6	38.9	41.9	45.6	48.3
27	11.8	12.9	14.6	16.2	18.1	21.7	26.3	31.5	36.7	40.1	43.2	47.0	49.6
28	12.5	13.6	15.3	16.9	18.9	22.7	27.3	32.6	37.9	41.3	44.5	48.3	51.0
29	13.1	14.3	16.0	17.7	19.8	23.6	28.3	33.7	39.1	42.6	45.7	49.6	52.3
30	13.8	15.0	16.8	18.5	20.6	24.5	29.3	34.8	40.3	43.8	47.0	50.9	53.7

For $n > 30$, the quantity $\sqrt{2\chi^2} - \sqrt{2n-1}$ may be treated as a critical ratio and referred to Table 3.

* This table is abridged from Catherine M. Thompson's "Table of Percentage Points of the χ^2 Distribution," copyright 1941 by *Biometrika*, **32**, 188–189, and used by permission of the author and editor.

TABLE 11   301

Table 11   Critical Values of the Pearson Correlation Coefficient*

df	Level of significance for one-tailed test			
	.05	.025	.01	.005
	Level of significance for two-tailed test			
	.10	.05	.02	.01
1	.988	.997	.9995	.9999
2	.900	.950	.980	.990
3	.805	.878	.934	.959
4	.729	.811	.882	.917
5	.669	.754	.833	.874
6	.622	.707	.789	.834
7	.582	.666	.750	.798
8	.549	.632	.716	.765
9	.521	.602	.685	.735
10	.497	.576	.658	.708
11	.576	.553	.634	.684
12	.458	.532	.612	.661
13	.441	.514	.592	.641
14	.426	.497	.574	.623
15	.412	.482	.558	.606
16	.400	.468	.542	.590
17	.389	.456	.528	.575
18	.378	.444	.516	.561
19	.369	.433	.503	.549
20	.360	.423	.492	.537
21	.352	.413	.482	.526
22	.344	.404	.472	.515
23	.337	.396	.462	.505
24	.330	.388	.453	.496
25	.323	.381	.445	.487
26	.317	.374	.437	.479
27	.311	.367	.430	.471
28	.306	.361	.423	.463
29	.301	.355	.416	.486
30	.296	.349	.409	.449
35	.275	.325	.381	.418
40	.257	.304	.358	.393
45	.243	.288	.338	.372
50	.231	.273	.322	.354
60	.211	.250	.295	.325
70	.195	.232	.274	.303
80	.183	.217	.256	.283
90	.173	.205	.242	.267
100	.164	.195	.230	.254

* Abridged from R. A. Fisher and F. Yates, *Statistical Tables for Biological, Agricultural, and Medical Research*, Oliver & Boyd, Ltd., Edinburgh, by permission of the authors and publishers.

Table 12   Critical Values of the Spearman Correlation Coefficient*

N	Significance level (one-tailed test)	
	.05	.01
4	1.000	
5	.900	1.000
6	.829	.943
7	.714	.893
8	.643	.833
9	.600	.783
10	.564	.746
12	.506	.712
14	.456	.645
16	.425	.601
18	.399	.564
20	.377	.534
22	.359	.508
24	.343	.485
26	.329	.465
28	.317	.448
30	.306	.432

* Adapted from E. G. Olds, "Distributions of sums of squares of rank differences for small numbers of individuals," *Annals of Mathematical Statistics*, **9**, 133–148, 1938; The 5% significance levels for sums of squares of rank differences and a correction, *Annals of Mathematical Statistics*, **20**, 117–118, 1949; with the kind permission of the author and the publisher.

TABLE 13   **303**

Table 13   Table of Critical Values of D in the Kolmogorov One-sample Test*

Sample size (N)	Level of significance for $D = maximum \|F_0(X) - S_N(X)\|$				
	.20	.15	.10	.05	.01
1	.900	.925	.950	.975	.995
2	.684	.726	.776	.842	.929
3	.565	.597	.642	.708	.828
4	.494	.525	.564	.624	.733
5	.446	.474	.510	.565	.669
6	.410	.436	.470	.521	.618
7	.381	.405	.438	.486	.577
8	.358	.381	.411	.457	.543
9	.339	.360	.388	.432	.514
10	.322	.342	.368	.410	.490
11	.307	.326	.352	.391	.468
12	.295	.313	.338	.375	.450
13	.284	.302	.325	.361	.433
14	.274	.292	.314	.349	.418
15	.266	.283	.304	.338	.404
16	.258	.274	.295	.328	.392
17	.250	.266	.286	.318	.381
18	.244	.259	.278	.309	.371
19	.237	.252	.272	.301	.363
20	.231	.246	.264	.294	.356
25	.21	.22	.24	.27	.32
30	.19	.20	.22	.24	.29
35	.18	.19	.21	.23	.27
Over 35	$\dfrac{1.07}{\sqrt{N}}$	$\dfrac{1.14}{\sqrt{N}}$	$\dfrac{1.22}{\sqrt{N}}$	$\dfrac{1.36}{\sqrt{N}}$	$\dfrac{1.63}{\sqrt{N}}$

* Adapted from F. J. Massey, Jr., "The Kolmogorov-Smirnov Test for Goodness of Fit," *J. Amer. Statist. Ass.*, **46**, 70, 1951, with the kind permission of the author and publisher.

Table 14   Table of Critical Values of K_D in the Kolmogorov-Smirnov Test for Two Samples of Equal Size
(*N is the size of each sample*)

N	One-tailed test*		Two-tailed test†	
	$\alpha = .05$	$\alpha = .01$	$\alpha = .05$	$\alpha = .01$
3	3	—	—	—
4	4	—	4	—
5	4	5	5	5
6	5	6	5	6
7	5	6	6	6
8	5	6	6	7
9	6	7	6	7
10	6	7	7	8
11	6	8	7	8
12	6	8	7	8
13	7	8	7	9
14	7	8	8	9
15	7	9	8	9
16	7	9	8	10
17	8	9	8	10
18	8	10	9	10
19	8	10	9	10
20	8	10	9	11
21	8	10	9	11
22	9	11	9	11
23	9	11	10	11
24	9	11	10	12
25	9	11	10	12
26	9	11	10	12
27	9	12	10	12
28	10	12	11	13
29	10	12	11	13
30	10	12	11	13
35	11	13	12	
40	11	14	13	

* Abridged from L. A. Goodman, "Kolomogorov-Smirnov tests for psychological research," *Psychol. Bull.*, **51**, 167, 1954, with the kind permission of the author and the American Psychological Association.

† Derived from Table 1 of F. J. Massey, Jr., "The distribution of the maximum deviation between two sample cumulative step functions," *Ann. Math. Statist.*, **22**, 126–127, 1951, with the kind permission of the author and the publisher.

TABLE 15   305

Table 15   Table of Critical Values of D in the Kolmogorov-Smirnov Test for Two Large Samples*

Level of significance	Value of D so large as to call for rejection of H_0 at the indicated level of significance, where $D = maximum \ |S_{n_1}(X) - S_{n_2}(X)|$
.10	$1.22 \sqrt{\dfrac{n_1 + n_2}{n_1 n_2}}$
.05	$1.36 \sqrt{\dfrac{n_1 + n_2}{n_1 n_2}}$
.025	$1.48 \sqrt{\dfrac{n_1 + n_2}{n_1 n_2}}$
.01	$1.63 \sqrt{\dfrac{n_1 + n_2}{n_1 n_2}}$
.005	$1.73 \sqrt{\dfrac{n_1 + n_2}{n_1 n_2}}$
.001	$1.95 \sqrt{\dfrac{n_1 + n_2}{n_1 n_2}}$

* Adapted from N. Smirnov, "Tables for estimating the goodness of fit of empirical distributions," *Ann. Math. Statist.*, **19**, 280–281, 1948, with the kind permission of the publisher.

Table 16   Table of Critical Values for the Fisher Test

Totals in right margin		B (or A)	Level of significance			
			.05	.025	.01	.005
$A + B = 3$	$C + D = 3$	3	0	—	—	—
$A + B = 4$	$C + D = 4$	4	0	0	—	—
	$C + D = 3$	4	0	—	—	—
$A + B = 5$	$C + D = 5$	5	1	1	0	0
		4	0	0	—	—
	$C + D = 4$	5	1	0	0	—
		4	0	—	—	—
	$C + D = 3$	5	0	0	—	—
	$C + D = 2$	5	0	—	—	—
$A + B = 6$	$C + D = 6$	6	2	1	1	0
		5	1	0	0	—
		4	0	—	—	—
	$C + D = 5$	6	1	0	0	0
		5	0	0	—	—
		4	0	—	—	—
	$C + D = 4$	6	1	0	0	0
		5	0	0	—	—
	$C + D = 3$	6	0	0	—	—
		5	0	—	—	—
	$C + D = 2$	6	0	—	—	—
$A + B = 7$	$C + D = 7$	7	3	2	1	1
		6	1	1	0	0
		5	0	0	—	—
		4	0	—	—	—
	$C + D = 6$	7	2	2	1	1
		6	1	0	0	0
		5	0	0	—	—
		4	0	—	—	—
	$C + D = 5$	7	2	1	0	0
		6	1	0	0	—
		5	0	—	—	—

NOTE: Critical values are maximum values of D (or C).

* Adapted from D. J. Finney, "The Fisher-Yates test of significance in 2 × 2 contingency tables," *Biometrika*, **35**, 149–154, 1948, with the kind permission of the author and the publisher.

TABLE 16 **307**

Table 16 Fisher Test (*Continued*)

Totals in right margin		B (or A)	Level of significance			
			.05	.025	.01	.005
A + B = 7	C + D = 4	7	1	1	0	0
		6	0	0	—	—
		5	0	—	—	—
	C + D = 3	7	0	0	0	—
		6	0	—	—	—
	C + D = 2	7	0	—	—	—
A + B = 8	C + D = 8	8	4	3	2	2
		7	2	2	1	0
		6	1	1	0	0
		5	0	0	—	—
		4	0	—	—	—
	C + D = 7	8	3	2	2	1
		7	2	1	1	0
		6	1	0	0	—
		5	0	0	—	—
	C + D = 6	8	2	2	1	1
		7	1	1	0	0
		6	0	0	0	—
		5	0	—	—	—
	C + D = 5	8	2	1	1	0
		7	1	0	0	0
		6	0	0	—	—
		5	0	—	—	—
	C + D = 4	8	1	1	0	0
		7	0	0	—	—
		6	0	—	—	—
	C + D = 3	8	0	0	0	—
		7	0	0	—	—
	C + D = 2	8	0	0	—	—
A + B = 9	C + D = 9	9	5	4	3	3
		8	3	3	2	1
		7	2	1	1	0
		6	1	1	0	0
		5	0	0	—	—
		4	0	—	—	—

NOTE: Critical values are maximum values of D (or C).

* When B is entered in the middle column, the significance levels are for D. When A is used in place of B, the significance levels are for C.

Table 16 Fisher Test (*Continued*)

Totals in right margin		B (or A)	Level of significance			
			.05	.025	.01	.005
$A + B = 9$	$C + D = 8$	9	4	3	3	2
		8	3	2	1	1
		7	2	1	0	0
		6	1	0	0	—
		5	0	0	—	—
	$C + D = 7$	9	3	3	2	2
		8	2	2	1	0
		7	1	1	0	0
		6	0	0	—	—
		5	0	—	—	—
	$C + D = 6$	9	3	2	1	1
		8	2	1	0	0
		7	1	0	0	—
		6	0	0	—	—
		5	0	—	—	—
	$C + D = 5$	9	2	1	1	1
		8	1	1	0	0
		7	0	0	—	—
		6	0	—	—	—
	$C + D = 4$	9	1	1	0	0
		8	0	0	0	—
		7	0	0	—	—
		6	0	—	—	—
	$C + D = 3$	9	1	0	0	0
		8	0	0	—	—
		7	0	—	—	—
	$C + D = 2$	9	0	0	—	—
$A + B = 10$	$C + D = 10$	10	6	5	4	3
		9	4	3	3	2
		8	3	2	1	1
		7	2	1	1	0
		6	1	0	0	—
		5	0	0	—	—
		4	0	—	—	—
	$C + D = 9$	10	5	4	3	3
		9	4	3	2	2

NOTE: Critical values are maximum values of D (or C).

TABLE 16   **309**

Table 16   **Fisher Test (*Continued*)**

Totals in right margin		B (or A)	Level of significance			
			.05	.025	.01	.005
$A + B = 10$ $C + D = 9$		8	2	2	1	1
		7	1	1	0	0
		6	1	0	0	—
		5	0	0	—	—
	$C + D = 8$	10	4	4	3	2
		9	3	2	2	1
		8	2	1	1	0
		7	1	1	0	0
		6	0	0	—	—
		5	0	—	—	—
	$C + D = 7$	10	3	3	2	2
		9	2	2	1	1
		8	1	1	0	0
		7	1	0	0	—
		6	0	0	—	—
		5	0	—	—	—
	$C + D = 6$	10	3	2	2	1
		9	2	1	1	0
		8	1	1	0	0
		7	0	0	—	—
		6	0	—	—	—
	$C + D = 5$	10	2	2	1	1
		9	1	1	0	0
		8	1	0	0	—
		7	0	0	—	—
		6	0	—	—	—
	$C + D = 4$	10	1	1	0	0
		9	1	0	0	0
		8	0	0	—	—
		7	0	—	—	—
	$C + D = 3$	10	1	0	0	0
		9	0	0	—	—
		8	0	—	—	—
	$C + D = 2$	10	0	0	—	—
		9	0	—	—	—
$A + B = 11$ $C + D = 11$		11	7	6	5	4
		10	5	4	3	3

NOTE: Critical values are maximum values of D (or C).

Table 16 Fisher Test (*Continued*)

Totals in right margin		B (or A)	Level of significance			
			.05	.025	.01	.005
$A + B = 11$	$C + D = 11$	9	4	3	2	2
		8	3	2	1	1
		7	2	1	0	0
		6	1	0	0	—
		5	0	0	—	—
		4	0	—	—	—
	$C + D = 10$	11	6	5	4	4
		10	4	4	3	2
		9	3	3	2	1
		8	2	2	1	0
		7	1	1	0	0
		6	1	0	0	—
		5	0	—	—	—
	$C + D = 9$	11	5	4	4	3
		10	4	3	2	2
		9	3	2	1	1
		8	2	1	1	0
		7	1	1	0	0
		6	0	0	—	—
		5	0	—	—	—
	$C + D = 8$	11	4	4	3	3
		10	3	3	2	1
		9	2	2	1	1
		8	1	1	0	0
		7	1	0	0	—
		6	0	0	—	—
		5	0	—	—	—
	$C + D = 7$	11	4	3	2	2
		10	3	2	1	1
		9	2	1	1	0
		8	1	1	0	0
		7	0	0	—	—
		6	0	0	—	—
	$C + D = 6$	11	3	2	2	1
		10	2	1	1	0
		9	1	1	0	0
		8	1	0	0	—
		7	0	0	—	—
		6	0	—	—	—

NOTE: Critical values are maximum values of D (or C).

TABLE 16   311

Table 16   Fisher Test (*Continued*)

Totals in right margin		B (or A)	Level of significance			
			.05	.025	.01	.005
A + B = 11	C + D = 5	11	2	2	1	1
		10	1	1	0	0
		9	1	0	0	0
		8	0	0	—	—
		7	0	—	—	—
	C + D = 4	11	1	1	1	0
		10	1	0	0	0
		9	0	0	—	—
		8	0	—	—	—
	C + D = 3	11	1	0	0	0
		10	0	0	—	—
		9	0	—	—	—
	C + D = 2	11	0	0	—	—
		10	0	—	—	—
A + B = 12	C + D = 12	12	8	7	6	5
		11	6	5	4	4
		10	5	4	3	2
		9	4	3	2	1
		8	3	2	1	1
		7	2	1	0	0
		6	1	0	0	—
		5	0	0	—	—
		4	0	—	—	—
	C + D = 11	12	7	6	5	5
		11	5	5	4	3
		10	4	3	2	2
		9	3	2	2	1
		8	2	1	1	0
		7	1	1	0	0
		6	1	0	0	—
		5	0	0	—	—
	C + D = 10	12	6	5	5	4
		11	5	4	3	3
		10	4	3	2	2
		9	3	2	1	1
		8	2	1	0	0
		7	1	0	0	0
		6	0	0	—	—
		5	0	—	—	—

NOTE: Critical values are maximum values of D (or C).

Table 16   Fisher Test (*Continued*)

Totals in right margin		B (or A)	Level of significance			
			.05	.025	.01	.005
$A + B = 12$ $C + D = 9$		12	5	5	4	3
		11	4	3	3	2
		10	3	2	2	1
		9	2	2	1	0
		8	1	1	0	0
		7	1	0	0	—
		6	0	0	—	—
		5	0	—	—	—
	$C + D = 8$	12	5	4	3	3
		11	3	3	2	2
		10	2	2	1	1
		9	2	1	1	0
		8	1	1	0	0
		7	0	0	—	—
		6	0	0	—	—
	$C + D = 7$	12	4	3	3	2
		11	3	2	2	1
		10	2	1	1	0
		9	1	1	0	0
		8	1	0	0	—
		7	0	0	—	—
		6	0	—	—	—
	$C + D = 6$	12	3	3	2	2
		11	2	2	1	1
		10	1	1	0	0
		9	1	0	0	0
		8	0	0	—	—
		7	0	0	—	—
		6	0	—	—	—
	$C + D = 5$	12	2	2	1	1
		11	1	1	1	0
		10	1	0	0	0
		9	0	0	0	—
		8	0	0	—	—
		7	0	—	—	—
	$C + D = 4$	12	2	1	1	0
		11	1	0	0	0
		10	0	0	0	—
		9	0	0	—	—
		8	0	—	—	—

NOTE: Critical values are maximum values of D (or C).

TABLE 16   313

Table 16   Fisher Test (*Continued*)

Totals in right margin		B (or A)	Level of significance			
			.05	.025	.01	.005
A + B = 12	C + D = 3	12	1	0	0	0
		11	0	0	0	—
		10	0	0	—	—
		9	0	—	—	—
	C + D = 2	12	0	0	—	—
		11	0	—	—	—
A + B = 13	C + D = 13	13	9	8	7	6
		12	7	6	5	4
		11	6	5	4	3
		10	4	4	3	2
		9	3	3	2	1
		8	2	2	1	0
		7	2	1	0	0
		6	1	0	0	—
		5	0	0	—	—
		4	0	—	—	—
	C + D = 12	13	8	7	6	5
		12	6	5	5	4
		11	5	4	3	3
		10	4	3	2	2
		9	3	2	1	1
		8	2	1	1	0
		7	1	1	0	0
		6	1	0	0	—
		5	0	0	—	—
	C + D = 11	13	7	6	5	5
		12	6	5	4	3
		11	4	4	3	2
		10	3	3	2	1
		9	3	2	1	1
		8	2	1	0	0
		7	1	0	0	0
		6	0	0	—	—
		5	0	—	—	—
	C + D = 10	13	6	6	5	4
		12	5	4	3	3
		11	4	3	2	2
		10	3	2	1	1

NOTE: Critical values are maximum values of *D* (or *C*).

Table 16   Fisher Test (*Continued*)

Totals in right margin		B (or A)	Level of significance			
			.05	.025	.01	.005
A + B = 13	C + D = 10	9	2	1	1	0
		8	1	1	0	0
		7	1	0	0	—
		6	0	0	—	—
		5	0	—	—	—
	C + D = 9	13	5	5	4	4
		12	4	4	3	2
		11	3	3	2	1
		10	2	2	1	1
		9	2	1	0	0
		8	1	1	0	0
		7	0	0	—	—
		6	0	0	—	—
		5	0	—	—	—
	C + D = 8	13	5	4	3	3
		12	4	3	2	2
		11	3	2	1	1
		10	2	1	1	0
		9	1	1	0	0
		8	1	0	0	—
		7	0	0	—	—
		6	0	—	—	—
	C + D = 7	13	4	3	3	2
		12	3	2	2	1
		11	2	2	1	1
		10	1	1	0	0
		9	1	0	0	0
		8	0	0	—	—
		7	0	0	—	—
		6	0	—	—	—
	C + D = 6	13	3	3	2	2
		12	2	2	1	1
		11	2	1	1	0
		10	1	1	0	0
		9	1	0	0	—
		8	0	0	—	—
		7	0	—	—	—
	C + D = 5	13	2	2	1	1
		12	2	1	1	0

NOTE: Critical values are maximum values of *D* (or *C*).

TABLE 16   315

Table 16   Fisher Test (*Continued*)

Totals in right margin		B (or A)	Level of significance			
			.05	.025	.01	.005
A + B = 13	C + D = 5	11	1	1	0	0
		10	1	0	0	—
		9	0	0	—	—
		8	0	—	—	—
	C + D = 4	13	2	1	1	0
		12	1	1	0	0
		11	0	0	0	—
		10	0	0	—	—
		9	0	—	—	—
	C + D = 3	13	1	1	0	0
		12	0	0	0	—
		11	0	0	—	—
		10	0	—	—	—
	C + D = 2	13	0	0	0	—
		12	0	—	—	—
A + B = 14	C + D = 14	14	10	9	8	7
		13	8	7	6	5
		12	6	6	5	4
		11	5	4	3	3
		10	4	3	2	2
		9	3	2	2	1
		8	2	2	1	0
		7	1	1	0	0
		6	1	0	0	—
		5	0	0	—	—
		4	0	—	—	—
	C + D = 13	14	9	8	7	6
		13	7	6	5	5
		12	6	5	4	3
		11	5	4	3	2
		10	4	3	2	2
		9	3	2	1	1
		8	2	1	1	0
		7	1	1	0	0
		6	1	0	—	—
		5	0	0	—	—
	C + D = 12	14	8	7	6	6
		13	6	6	5	4

NOTE: Critical values are maximum values of D (or C).

Table 16   Fisher Test (*Continued*)

Totals in right margin		B (or A)	Level of significance			
			.05	.025	.01	.005
A + B = 14	C + D = 12	12	5	4	4	3
		11	4	3	3	2
		10	3	3	2	1
		9	2	2	1	1
		8	2	1	0	0
		7	1	0	0	—
		6	0	0	—	—
		5	0	—	—	—
	C + D = 11	14	7	6	6	5
		13	6	5	4	4
		12	5	4	3	3
		11	4	3	2	2
		10	3	2	1	1
		9	2	1	1	0
		8	1	1	0	0
		7	1	0	0	—
		6	0	0	—	—
		5	0	—	—	—
	C + D = 10	14	6	6	5	4
		13	5	4	4	3
		12	4	3	3	2
		11	3	3	2	1
		10	2	2	1	1
		9	2	1	0	0
		8	1	1	0	0
		7	0	0	0	—
		6	0	0	—	—
		5	0	—	—	—
	C + D = 9	14	6	5	4	4
		13	4	4	3	3
		12	3	3	2	2
		11	3	2	1	1
		10	2	1	1	0
		9	1	1	0	0
		8	1	0	0	—
		7	0	0	—	—
		6	0	—	—	—
	C + D = 8	14	5	4	4	3
		13	4	3	2	2

NOTE: Critical values are maximum values of *D* (or *C*).

TABLE 16   317

Table 16   Fisher Test (*Continued*)

Totals in right margin		B (or A)	Level of significance			
			.05	.025	.01	.005
A + B = 14	C + D = 8	12	3	2	2	1
		11	2	2	1	1
		10	2	1	0	0
		9	1	0	0	0
		8	0	0	0	—
		7	0	0	—	—
		6	0	—	—	—
	C + D = 7	14	4	3	3	2
		13	3	2	2	1
		12	2	2	1	1
		11	2	1	1	0
		10	1	1	0	0
		9	1	0	0	—
		8	0	0	—	—
		7	0	—	—	—
	C + D = 6	14	3	3	2	2
		13	2	2	1	1
		12	2	1	1	0
		11	1	1	0	0
		10	1	0	0	—
		9	0	0	—	—
		8	0	0	—	—
		7	0	—	—	—
	C + D = 5	14	2	2	1	1
		13	2	1	1	0
		12	1	1	0	0
		11	1	0	0	0
		10	0	0	—	—
		9	0	0	—	—
		8	0	—	—	—
	C + D = 4	14	2	1	1	1
		13	1	1	0	0
		12	1	0	0	0
		11	0	0	—	—
		10	0	0	—	—
		9	0	—	—	—
	C + D = 3	14	1	1	0	0
		13	0	0	0	—
		12	0	0	—	—
		11	0	—	—	—

NOTE: Critical values are maximum values of D (or C).

Table 16   Fisher Test (*Continued*)

Totals in right margin	B (or A)	Level of significance			
		.05	.025	.01	.005
$A + B = 14$ $C + D = 2$	14	0	0	0	—
	13	0	0	—	—
	12	0	—	—	—
$A + B = 15$ $C + D = 15$	15	11	10	9	8
	14	9	8	7	6
	13	7	6	5	5
	12	6	5	4	4
	11	5	4	3	3
	10	4	3	2	2
	9	3	2	1	1
	8	2	1	1	0
	7	1	1	0	0
	6	1	0	0	—
	5	0	0	—	—
	4	0	—	—	—
$C + D = 14$	15	10	9	8	7
	14	8	7	6	6
	13	7	6	5	4
	12	6	5	4	3
	11	5	4	3	2
	10	4	3	2	1
	9	3	2	1	1
	8	2	1	1	0
	7	1	1	0	0
	6	1	0	—	—
	5	0	—	—	—
$C + D = 13$	15	9	8	7	7
	14	7	7	6	5
	13	6	5	4	4
	12	5	4	3	3
	11	4	3	2	2
	10	3	2	2	1
	9	2	2	1	0
	8	2	1	0	0
	7	1	0	0	—
	6	0	0	—	—
	5	0	—	—	—

NOTE: Critical values are maximum values of D (or C).

TABLE 16   319

Table 16   Fisher Test (*Continued*)

Totals in right margin		B (or A)	Level of significance			
			.05	.025	.01	.005
A + B = 15	C + D = 12	15	8	7	7	6
		14	7	6	5	4
		13	6	5	4	3
		12	5	4	3	2
		11	4	3	2	2
		10	3	2	1	1
		9	2	1	1	0
		8	1	1	0	0
		7	1	0	0	—
		6	0	0	—	—
		5	0	—	—	—
	C + D = 11	15	7	7	6	5
		14	6	5	4	4
		13	5	4	3	3
		12	4	3	2	2
		11	3	2	2	1
		10	2	2	1	1
		9	2	1	0	0
		8	1	1	0	0
		7	1	0	0	—
		6	0	0	—	—
		5	0	—	—	—
	C + D = 10	15	6	6	5	5
		14	5	5	4	3
		13	4	4	3	2
		12	3	3	2	2
		11	3	2	1	1
		10	2	1	1	0
		9	1	1	0	0
		8	1	0	0	—
		7	0	0	—	—
		6	0	—	—	—
	C + D = 9	15	6	5	4	4
		14	5	4	3	3
		13	4	3	2	2
		12	3	2	2	1
		11	2	2	1	1
		10	2	1	0	0
		9	1	1	0	0

NOTE: Critical values are maximum values of D (or C).

Table 16   Fisher Test (*Continued*)

Totals in right margin		B (or A)	Level of significance			
			.05	.025	.01	.005
A + B = 15	C + D = 9	8	1	0	0	—
		7	0	0	—	—
		6	0	—	—	—
	C + D = 8	15	5	4	4	3
		14	4	3	3	2
		13	3	2	2	1
		12	2	2	1	1
		11	2	1	1	0
		10	1	1	0	0
		9	1	0	0	—
		8	0	0	—	—
		7	0	—	—	—
		6	0	—	—	—
	C + D = 7	15	4	4	3	3
		14	3	3	2	2
		13	2	2	1	1
		12	2	1	1	0
		11	1	1	0	0
		10	1	0	0	0
		9	0	0	—	—
		8	0	0	—	—
		7	0	—	—	—
	C + D = 6	15	3	3	2	2
		14	2	2	1	1
		13	2	1	1	0
		12	1	1	0	0
		11	1	0	0	0
		10	0	0	0	—
		9	0	0	—	—
		8	0	—	—	—
	C + D = 5	15	2	2	2	1
		14	2	1	1	1
		13	1	1	0	0
		12	1	0	0	0
		11	0	0	0	—
		10	0	0	—	—
		9	0	—	—	—
	C + D = 4	15	2	1	1	1
		14	1	1	0	0

NOTE: Critical values are maximum values of *D* (or *C*).

TABLE 16   321

Table 16   Fisher Test (*Continued*)

Totals in right margin		B (or A)	Level of significance			
			.05	.025	.01	.005
A + B = 15	C + D = 4	13	1	0	0	0
		12	0	0	0	—
		11	0	0	—	—
		10	0	—	—	—
	C + D = 3	15	1	1	0	0
		14	0	0	0	0
		13	0	0	—	—
		12	0	0	—	—
		11	0	—	—	—
	C + D = 2	15	0	0	0	—
		14	0	0	—	—
		13	0	—	—	—

NOTE: Critical values are maximum values of D (or C).

Table 17   The F-Distribution* (.05 Level)

one tail

df_2 \ df_1	1	2	3	4	5	6	7	8	9	10	12	15	20	24	30	40	60	120	∞
1	161.4	199.5	215.7	224.6	230.2	234.0	236.8	238.9	240.5	241.9	243.9	245.9	248.0	249.1	250.1	251.1	252.2	253.3	254.3
2	18.51	19.00	19.16	19.25	19.30	19.33	19.35	19.37	19.38	19.40	19.41	19.43	19.45	19.45	19.46	19.47	19.48	19.49	19.50
3	10.13	9.55	9.28	9.12	9.01	8.94	8.89	8.85	8.81	8.79	8.74	8.70	8.66	8.64	8.62	8.59	8.57	8.55	8.53
4	7.71	6.94	6.59	6.39	6.26	6.16	6.09	6.04	6.00	5.96	5.91	5.86	5.80	5.77	5.75	5.72	5.69	5.66	5.63
5	6.61	5.79	5.41	5.19	5.05	4.95	4.88	4.82	4.77	4.74	4.68	4.62	4.56	4.53	4.50	4.46	4.43	4.40	4.36
6	5.99	5.14	4.76	4.53	4.39	4.28	4.21	4.15	4.10	4.06	4.00	3.94	3.87	3.84	3.81	3.77	3.74	3.70	3.67
7	5.59	4.74	4.35	4.12	3.97	3.87	3.79	3.73	3.68	3.64	3.57	3.51	3.44	3.41	3.38	3.34	3.30	3.27	3.23
8	5.32	4.46	4.07	3.84	3.69	3.58	3.50	3.44	3.39	3.35	3.28	3.22	3.15	3.12	3.08	3.04	3.01	2.97	2.93
9	5.12	4.26	3.86	3.63	3.48	3.37	3.29	3.23	3.18	3.14	3.07	3.01	2.94	2.90	2.86	2.83	2.79	2.75	2.71
10	4.96	4.10	3.71	3.48	3.33	3.22	3.14	3.07	3.02	2.98	2.91	2.85	2.77	2.74	2.70	2.66	2.62	2.58	2.54
11	4.84	3.98	3.59	3.36	3.20	3.09	3.01	2.95	2.90	2.85	2.79	2.72	2.65	2.61	2.57	2.53	2.49	2.45	2.40
12	4.75	3.89	3.49	3.26	3.11	3.00	2.91	2.85	2.80	2.75	2.69	2.62	2.54	2.51	2.47	2.43	2.38	2.34	2.30
13	4.67	3.81	3.41	3.18	3.03	2.92	2.83	2.77	2.71	2.67	2.60	2.53	2.46	2.42	2.38	2.34	2.30	2.25	2.21
14	4.60	3.74	3.34	3.11	2.96	2.85	2.76	2.70	2.65	2.60	2.53	2.46	2.39	2.35	2.31	2.27	2.22	2.18	2.13
15	4.54	3.68	3.29	3.06	2.90	2.79	2.71	2.64	2.59	2.54	2.48	2.40	2.33	2.29	2.25	2.20	2.16	2.11	2.07
16	4.49	3.63	3.24	3.01	2.85	2.74	2.66	2.59	2.54	2.49	2.42	2.35	2.28	2.24	2.19	2.15	2.11	2.06	2.01
17	4.45	3.59	3.20	2.96	2.81	2.70	2.61	2.55	2.49	2.45	2.38	2.31	2.23	2.19	2.15	2.10	2.06	2.01	1.96
18	4.41	3.55	3.16	2.93	2.77	2.66	2.58	2.51	2.46	2.41	2.34	2.27	2.19	2.15	2.11	2.06	2.02	1.97	1.92
19	4.38	3.52	3.13	2.90	2.74	2.63	2.54	2.48	2.42	2.38	2.31	2.23	2.16	2.11	2.07	2.03	1.98	1.93	1.88
20	4.35	3.49	3.10	2.87	2.71	2.60	2.51	2.45	2.39	2.35	2.28	2.20	2.12	2.08	2.04	1.99	1.95	1.90	1.84
21	4.32	3.47	3.07	2.84	2.68	2.57	2.49	2.42	2.37	2.32	2.25	2.18	2.10	2.05	2.01	1.96	1.92	1.87	1.81
22	4.30	3.44	3.05	2.82	2.66	2.55	2.46	2.40	2.34	2.30	2.23	2.15	2.07	2.03	1.98	1.94	1.89	1.84	1.78
23	4.28	3.42	3.03	2.80	2.64	2.53	2.44	2.37	2.32	2.27	2.20	2.13	2.05	2.01	1.96	1.91	1.86	1.81	1.76
24	4.26	3.40	3.01	2.78	2.62	2.51	2.42	2.36	2.30	2.25	2.18	2.11	2.03	1.98	1.94	1.89	1.84	1.79	1.73
25	4.24	3.39	2.99	2.76	2.60	2.49	2.40	2.34	2.28	2.24	2.16	2.09	2.01	1.96	1.92	1.87	1.82	1.77	1.71
26	4.23	3.37	2.98	2.74	2.59	2.47	2.39	2.32	2.27	2.22	2.15	2.07	1.99	1.95	1.90	1.85	1.80	1.75	1.69
27	4.21	3.35	2.96	2.73	2.57	2.46	2.37	2.31	2.25	2.20	2.13	2.06	1.97	1.93	1.88	1.84	1.79	1.73	1.67
28	4.20	3.34	2.95	2.71	2.56	2.45	2.36	2.29	2.24	2.19	2.12	2.04	1.96	1.91	1.87	1.82	1.77	1.71	1.65
29	4.18	3.33	2.93	2.70	2.55	2.43	2.35	2.28	2.22	2.18	2.10	2.03	1.94	1.90	1.85	1.81	1.75	1.70	1.64
30	4.17	3.32	2.92	2.69	2.53	2.42	2.33	2.27	2.21	2.16	2.09	2.01	1.93	1.89	1.84	1.79	1.74	1.68	1.62
40	4.08	3.23	2.84	2.61	2.45	2.34	2.25	2.18	2.12	2.08	2.00	1.92	1.84	1.79	1.74	1.69	1.64	1.58	1.51
60	4.00	3.15	2.76	2.53	2.37	2.25	2.17	2.10	2.04	1.99	1.92	1.84	1.75	1.70	1.65	1.59	1.53	1.47	1.39
120	3.92	3.07	2.68	2.45	2.29	2.17	2.09	2.02	1.96	1.91	1.83	1.75	1.66	1.61	1.55	1.50	1.43	1.35	1.25
∞	3.84	3.00	2.60	2.37	2.21	2.10	2.01	1.94	1.88	1.83	1.75	1.67	1.57	1.52	1.46	1.39	1.32	1.22	1.00

TABLE 17   323

Table 17   The F-Distribution* (.01 Level) (Continued)

df_2 \ df_1	1	2	3	4	5	6	7	8	9	10	12	15	20	24	30	40	60	120	∞
1	4052	4999.5	5403	5625	5764	5859	5928	5982	6022	6056	6106	6157	6209	6235	6261	6287	6313	6339	6366
2	98.5	99.00	99.17	99.25	99.30	99.33	99.36	99.37	99.39	99.40	99.42	99.43	99.45	99.46	99.47	99.47	99.48	99.49	99.50
3	34.12	30.82	29.46	28.71	28.24	27.91	27.67	27.49	27.35	27.23	27.05	26.87	26.69	26.60	26.50	26.41	26.32	26.22	26.13
4	21.20	18.00	16.69	15.98	15.52	15.21	14.98	14.80	14.66	14.55	14.37	14.20	14.02	13.93	13.84	13.75	13.65	13.56	13.46
5	16.26	13.27	12.06	11.39	10.97	10.67	10.46	10.29	10.16	10.05	9.89	9.72	9.55	9.47	9.38	9.29	9.20	9.11	9.02
6	13.75	10.92	9.78	9.15	8.75	8.47	8.26	8.10	7.98	7.87	7.72	7.56	7.40	7.31	7.23	7.14	7.06	6.97	6.88
7	12.25	9.55	8.45	7.85	7.46	7.19	6.99	6.84	6.72	6.62	6.47	6.31	6.16	6.07	5.99	5.91	5.82	5.74	5.65
8	11.26	8.65	7.59	7.01	6.63	6.37	6.18	6.03	5.91	5.81	5.67	5.52	5.36	5.28	5.20	5.12	5.03	4.95	4.86
9	10.56	8.02	6.99	6.42	6.06	5.80	5.61	5.47	5.35	5.26	5.11	4.96	4.81	4.73	4.65	4.57	4.48	4.40	4.31
10	10.04	7.56	6.55	5.99	5.64	5.39	5.20	5.06	4.94	4.85	4.71	4.56	4.41	4.33	4.25	4.17	4.08	4.00	3.91
11	9.65	7.21	6.22	5.67	5.32	5.07	4.89	4.74	4.63	4.54	4.40	4.25	4.10	4.02	3.94	3.86	3.78	3.69	3.60
12	9.33	6.93	5.95	5.41	5.06	4.82	4.64	4.50	4.39	4.30	4.16	4.01	3.86	3.78	3.70	3.62	3.54	3.45	3.36
13	9.07	6.70	5.74	5.21	4.86	4.62	4.44	4.30	4.19	4.10	3.96	3.82	3.66	3.59	3.51	3.43	3.34	3.25	3.17
14	8.86	6.51	5.56	5.04	4.69	4.46	4.28	4.14	4.03	3.94	3.80	3.66	3.51	3.43	3.35	3.27	3.18	3.09	3.00
15	8.68	6.36	5.42	4.89	4.56	4.32	4.14	4.00	3.89	3.80	3.67	3.52	3.37	3.29	3.21	3.13	3.05	2.96	2.87
16	8.53	6.23	5.29	4.77	4.44	4.20	4.03	3.89	3.78	3.69	3.55	3.41	3.26	3.18	3.10	3.02	2.93	2.84	2.75
17	8.40	6.11	5.18	4.67	4.34	4.10	3.93	3.79	3.68	3.59	3.46	3.31	3.16	3.08	3.00	2.92	2.83	2.75	2.65
18	8.29	6.01	5.09	4.58	4.25	4.01	3.84	3.71	3.60	3.51	3.37	3.23	3.08	3.00	2.92	2.84	2.75	2.66	2.57
19	8.18	5.93	5.01	4.50	4.17	3.94	3.77	3.63	3.52	3.43	3.30	3.15	3.00	2.92	2.84	2.76	2.67	2.58	2.49
20	8.10	5.85	4.94	4.43	4.10	3.87	3.70	3.56	3.46	3.37	3.23	3.09	2.94	2.86	2.78	2.69	2.61	2.52	2.42
21	8.02	5.78	4.87	4.37	4.04	3.81	3.64	3.51	3.40	3.31	3.17	3.03	2.88	2.80	2.72	2.64	2.55	2.46	2.36
22	7.95	5.72	4.82	4.31	3.99	3.76	3.59	3.45	3.35	3.26	3.12	2.98	2.83	2.75	2.67	2.58	2.50	2.40	2.31
23	7.88	5.66	4.76	4.26	3.94	3.71	3.54	3.41	3.30	3.21	3.07	2.93	2.78	2.70	2.62	2.54	2.45	2.35	2.26
24	7.82	5.61	4.72	4.22	3.90	3.67	3.50	3.36	3.26	3.17	3.03	2.89	2.74	2.66	2.58	2.49	2.40	2.31	2.21
25	7.77	5.57	4.68	4.18	3.85	3.63	3.46	3.32	3.22	3.13	2.99	2.85	2.70	2.62	2.54	2.45	2.36	2.27	2.17
26	7.72	5.53	4.64	4.14	3.82	3.59	3.42	3.29	3.18	3.09	2.96	2.81	2.66	2.58	2.50	2.42	2.33	2.23	2.13
27	7.68	5.49	4.60	4.11	3.78	3.56	3.39	3.26	3.15	3.06	2.93	2.78	2.63	2.55	2.47	2.38	2.29	2.20	2.10
28	7.64	5.45	4.57	4.07	3.75	3.53	3.36	3.23	3.12	3.03	2.90	2.75	2.60	2.52	2.44	2.35	2.26	2.17	2.06
29	7.60	5.42	4.54	4.04	3.73	3.50	3.33	3.20	3.09	3.00	2.87	2.73	2.57	2.49	2.41	2.33	2.23	2.14	2.03
30	7.56	5.39	4.51	4.02	3.70	3.47	3.30	3.17	3.07	2.98	2.84	2.70	2.55	2.47	2.39	2.30	2.21	2.11	2.01
40	7.31	5.18	4.31	3.83	3.51	3.29	3.12	2.99	2.89	2.80	2.66	2.52	2.37	2.29	2.20	2.11	2.02	1.92	1.80
60	7.08	4.98	4.13	3.65	3.34	3.12	2.95	2.82	2.72	2.63	2.50	2.35	2.20	2.12	2.03	1.94	1.84	1.73	1.60
120	6.85	4.79	3.95	3.48	3.17	2.96	2.79	2.66	2.56	2.47	2.34	2.19	2.03	1.95	1.86	1.76	1.66	1.53	1.38
∞	6.63	4.61	3.78	3.32	3.02	2.80	2.64	2.51	2.41	2.32	2.18	2.04	1.88	1.79	1.70	1.59	1.47	1.32	1.00

Table 18   The Dunnett Test* (.05 Level, One-tailed)

Number of treatment means (excluding the control)

df	1	2	3	4	5	6	7	8	9
5	2.02	2.44	2.68	2.85	2.98	3.08	3.16	3.24	3.30
6	1.94	2.34	2.56	2.71	2.83	2.92	3.00	3.07	3.12
7	1.89	2.27	2.48	2.62	2.73	2.82	2.89	2.95	3.01
8	1.86	2.22	2.42	2.55	2.66	2.74	2.81	2.87	2.92
9	1.83	2.18	2.37	2.50	2.60	2.68	2.75	2.81	2.86
10	1.81	2.15	2.34	2.47	2.56	2.64	2.70	2.76	2.81
11	1.80	2.13	2.31	2.44	2.53	2.60	2.67	2.72	2.77
12	1.78	2.11	2.29	2.41	2.50	2.58	2.64	2.69	2.74
13	1.77	2.09	2.27	2.39	2.48	2.55	2.61	2.66	2.71
14	1.76	2.08	2.25	2.37	2.46	2.53	2.59	2.64	2.69
15	1.75	2.07	2.24	2.36	2.44	2.51	2.57	2.62	2.67
16	1.75	2.06	2.23	2.34	2.43	2.50	2.56	2.61	2.65
17	1.74	2.05	2.22	2.33	2.42	2.49	2.54	2.59	2.64
18	1.73	2.04	2.21	2.32	2.41	2.48	2.53	2.58	2.62
19	1.73	2.03	2.20	2.31	2.40	2.47	2.52	2.57	2.61
20	1.72	2.03	2.19	2.30	2.39	2.46	2.51	2.56	2.60
24	1.71	2.01	2.17	2.28	2.36	2.43	2.48	2.53	2.57
30	1.70	1.99	2.15	2.25	2.33	2.40	2.45	2.50	2.54
40	1.68	1.97	2.13	2.23	2.31	2.37	2.42	2.47	2.51
60	1.67	1.95	2.10	2.21	2.28	2.35	2.39	2.44	2.48
120	1.66	1.93	2.08	2.18	2.26	2.32	2.37	2.41	2.45
inf.	1.64	1.92	2.06	2.16	2.23	2.29	2.34	2.38	2.42

* This table is reprinted from C. W. Dunnett, "A multiple comparison procedure for comparing several treatments with a control," *J. Amer. Statist. Ass.*, **50**, 1096–1121, 1955, by permission of the author and the editors of the *Journal of the American Statistical Association.*

TABLE 18   325

Table 18   The Dunnett Test (.05 Level, Two-tailed)

Number of treatment means (excluding the control)

df	1	2	3	4	5	6	7	8	9
5	2.57	3.03	3.39	3.66	3.88	4.06	4.22	4.36	4.49
6	2.45	2.86	3.18	3.41	3.60	3.75	3.88	4.00	4.11
7	2.36	2.75	3.04	3.24	3.41	3.54	3.66	3.76	3.86
8	2.31	2.67	2.94	3.13	3.28	3.40	3.51	3.60	3.68
9	2.26	2.61	2.86	3.04	3.18	3.29	3.39	3.48	3.55
10	2.23	2.57	2.81	2.97	3.11	3.21	3.31	3.39	3.46
11	2.20	2.53	2.76	2.92	3.05	3.15	3.24	3.31	3.38
12	2.18	2.50	2.72	2.88	3.00	3.10	3.18	3.25	3.32
13	2.16	2.48	2.69	2.84	2.96	3.06	3.14	3.21	3.27
14	2.14	2.46	2.67	2.81	2.93	3.02	3.10	3.17	3.23
15	2.13	2.44	2.64	2.79	2.90	2.99	3.07	3.13	3.19
16	2.12	2.42	2.63	2.77	2.88	2.96	3.04	3.10	3.16
17	2.11	2.41	2.61	2.75	2.85	2.94	3.01	3.08	3.13
17	2.10	2.40	2.59	2.73	2.84	2.92	2.99	3.05	3.11
19	2.09	2.39	2.58	2.72	2.82	2.90	2.97	3.04	3.09
20	2.09	2.38	2.57	2.70	2.81	2.89	2.96	3.02	3.07
24	2.06	2.35	2.53	2.66	2.76	2.84	2.91	2.96	3.01
30	2.04	2.32	2.50	2.62	2.72	2.79	2.86	2.91	2.96
40	2.02	2.29	2.47	2.58	2.67	2.75	2.81	2.86	2.90
60	2.00	2.27	2.43	2.55	2.63	2.70	2.76	2.81	2.85
120	1.98	2.24	2.40	2.51	2.59	2.66	2.71	2.76	2.80
inf.	1.96	2.21	2.37	2.47	2.55	2.62	2.67	2.71	2.75

Table 18   The Dunnett Test (.01 Level, One-tailed)

Number of treatment means (excluding the control)

df	1	2	3	4	5	6	7	8	9
5	3.37	3.90	4.21	4.43	4.60	4.73	4.85	4.94	5.03
6	3.14	3.61	3.88	4.07	4.21	4.33	4.43	4.51	4.59
7	3.00	3.42	3.66	3.83	3.96	4.07	4.15	4.23	4.30
8	2.90	3.29	3.51	3.67	3.79	3.88	3.96	4.03	4.09
9	2.82	3.19	3.40	3.55	3.66	3.75	3.82	3.89	3.94
10	2.76	3.11	3.31	3.45	3.56	3.64	3.71	3.78	3.83
11	2.72	3.06	3.25	3.38	3.48	3.56	3.63	3.69	3.74
12	2.68	3.01	3.19	3.32	3.42	3.50	3.56	3.62	3.67
13	2.65	2.97	3.15	3.27	3.37	3.44	3.51	3.56	3.61
14	2.62	2.94	3.11	3.23	3.32	3.40	3.46	3.51	3.56
15	2.60	2.91	3.08	3.20	3.29	3.36	3.42	3.47	3.52
16	2.58	2.88	3.05	3.17	3.26	3.33	3.39	3.44	3.48
17	2.57	2.86	3.03	3.14	3.23	3.30	3.36	3.41	3.45
18	2.55	2.84	3.01	3.12	3.21	3.27	3.33	3.38	3.42
19	2.54	2.83	2.99	3.10	3.18	3.25	3.31	3.36	3.40
20	2.53	2.81	2.97	3.08	3.17	3.23	3.29	3.34	3.38
24	2.49	2.77	2.92	3.03	3.11	3.17	3.22	3.27	3.31
30	2.46	2.72	2.87	2.97	3.05	3.11	3.16	3.21	3.24
40	2.42	2.68	2.82	2.92	2.99	3.05	3.10	3.14	3.18
60	2.39	2.64	2.78	2.87	2.94	3.00	3.04	3.08	3.12
120	2.36	2.60	2.73	2.82	2.89	2.94	2.99	3.03	3.06
inf.	2.33	2.56	2.68	2.77	2.84	2.89	2.93	2.97	3.00

TABLE 18   327

Table 18   The Dunnett Test (.01 Level, Two-tailed)

Number of treatment means (excluding the control)

df	1	2	3	4	5	6	7	8	9
5	4.03	4.63	5.09	5.44	5.73	5.97	6.18	6.36	6.53
6	3.71	4.22	4.60	4.88	5.11	5.30	5.47	5.61	5.74
7	3.50	3.95	4.28	4.52	4.71	4.87	5.01	5.13	5.24
8	3.36	3.77	4.06	4.27	4.44	4.58	4.70	4.81	4.90
9	3.25	3.63	3.90	4.09	4.24	4.37	4.48	4.57	4.65
10	3.17	3.53	3.78	3.95	4.10	4.21	4.31	4.40	4.47
11	3.11	3.45	3.68	3.85	3.98	4.09	4.18	4.26	4.33
12	3.05	3.39	3.61	3.76	3.89	3.99	4.08	4.15	4.22
13	3.01	3.33	3.54	3.69	3.81	3.91	3.99	4.06	4.13
14	2.98	3.29	3.49	3.64	3.75	3.84	3.92	3.99	4.05
15	2.95	3.25	3.45	3.59	3.70	3.79	3.86	3.93	3.99
16	2.92	3.22	3.41	3.55	3.65	3.74	3.82	3.88	3.93
17	2.90	3.19	3.38	3.51	3.62	3.70	3.77	3.83	3.89
18	2.88	3.17	3.35	3.48	3.58	3.67	3.74	3.80	3.85
19	2.86	3.15	3.33	3.46	3.55	3.64	3.70	3.76	3.81
20	2.85	3.13	3.31	3.43	3.53	3.61	3.67	3.73	3.78
24	2.80	3.07	3.24	3.36	3.45	3.52	3.58	3.64	3.69
30	2.75	3.01	3.17	3.28	3.37	3.44	3.50	3.55	3.59
40	2.70	2.95	3.10	3.21	3.29	3.36	3.41	3.46	3.50
60	2.66	2.90	3.04	3.14	3.22	3.28	3.33	3.38	3.42
120	2.62	2.84	2.98	3.08	3.15	3.21	3.25	3.30	3.33
inf.	2.58	2.79	2.92	3.01	3.08	3.14	3.18	3.22	3.25

ANSWERS TO SELECTED EXERCISES

3.2 (a)

X	f	fX	cf	rf	rcf
10	1	10	20	.05	1.00
9	2	18	19	.10	.95
8	3	24	17	.15	.85
7	1	7	14	.05	.70
6	4	24	13	.20	.65
5	6	30	9	.30	.45
4	2	8	3	.10	.15
3	1	3	1	.05	.05
	20	124		1.00	

(b) $N = 20$, $T = 124$. For answers to (c), (d), and (e), see (a) above.

4.1 $Q_1 = 90.5$; $Q_2 = 99.5$; $Q_3 = 108.5$.

4.2 $D_2 = 4.67$; $D_4 = 5.33$; $D_6 = 6.25$; $D_8 = 8.17$.

6.1 (a) 50; (b) 340; (c) 2500; (d) 90.

6.2 (a) 647; (b) 16.1; (c) 1,792.8; (d) 10,416.7; (e) 56.7.

6.3 (a) 30; (b) 16; (c) 75; (d) 285; (e) 5,625.

7.1 (a) 6; (b) 5.5; (c) 5; (d) 30; (e) 16; (f) 0.

7.2 (a) 12.03; (b) 11.3; (c) 10.

7.3 (a) 3.00; 2.50; 2.29; 2.38; (b) 2.50; (c) 2.50.

8.1 (a) 9; (b) 2.88; (c) 2.6; (d) 90; (e) 9; (f) 3; (g) 9; (h) 36; (i) 30.

8.2 (a) 98.92; (b) 8.24; (c) 2.87.

8.3 (a) 1912.59; (b) 15.94; (c) 3.99.

9.1 (a) 5; 2; (b) -2.0; $-.5$; $-.5$; 0; 0; .5; 1.0; 1.5; (c) 30; 45; 45; 50; 50; 55; 60; 65.

9.2 (a) 10; 2; (b) -1.5; -1.5; -1.0; $-.5$; $-.5$; 0; 0; .5; .5; 1.0; 1.5; 1.5; (c) 77.5; 77.5; 85; 92.5; 92.5; 100; 100; 107.5; 107.5; 115; 122.5; 122.5.

9.3   (a) 11.06; 3.99; (b) -2.52; -2.27; -2.02; -1.77; -1.52; -1.27; -1.02; $-.77$; $-.52$; $-.27$; $-.02$; .24; .49; .74; .99; 1.24; 1.49; 1.74; 1.99; 2.24.

10.1   (a) .4429; (b) .9429; (c) .0571; (d) 94.

10.2   (a) .4429; (b) .0571; (c) .9429; (d) 6.

10.3   (a) .4975; (b) .9975; (c) .0025; (d) 99.

10.4   (a) .4975; (b) .0025; (c) .9975; (d) 0+.

10.5 $-.7721$.

10.6   1.1341.

10.7   0.

10.8   (a) .0013; (b) .9987.

10.9   1.645.

10.10 $-.6745$ to $+.6745$.

11.1 -2.65; -2.41; -2.12; -1.79; -1.44; -1.19; $-.97$; $-.74$; $-.49$; $-.27$; $-.03$; .23; .48; .70; .90; 1.13; 1.44; 1.77; 2.12; 2.65.

11.2   (a) For IQ of 80, $z = -1.37$; for 90, $z = -.69$; for 100, $z = .09$; for 110, $z = .71$; for 120, $z = 1.41$; (b) For IQ of 80, $T = 36$, for 90, $T = 43$, for 100, $T = 51$; for 110, $T = 57$; for 120, $T = 64$; (c) For IQ of 80, 2; for 90, 4; for 100, 5; for 110, 6; for 120, 8.

12.1   (b) .517.

12.2   (b) $-.792$.

12.3   (a) $-.535$; (b) Girls scored higher than boys; (c) .287.

13.1   .70.

13.2   .612.

13.3 $-.522$.

13.4 $-.654$.

13.5   .351

13.6   .53.

14.1   (b) .90; (c) .09; (d) -7; (e) $Y = .09X - 7$; (g) 3.8; 2.9; 2.0; 1.1; .2; (h) 1.90; (i) 2; (k) .796; (l) 1.55.

17.1   (a) 15/67 or .224; (b) 27/67 or .403.

17.2   (a) 1/2 or .500; (b) 1/256 or .004; (c) 1/256 or .004.

17.3   (a) 1/3 or .333, (b) 2/3 or .667, (c) 1/243 or .004.

17.4   (a) 24; (b) 1/24 or .004.

17.5   35.

17.6   (a) 336; (b) 1/8 or .125; 1/336 or .003.

18.1   (a) 1/64 or .016; (b) 1/64 or .016; (c) 6/64 or .094; (d) 7/64 or .109.

18.2   (a) 1/729 or .0014; (b) 64/729 or .0878; (c) 12/729 or .0165; (d) 13/729 or .0178.

18.3   201/59,049 or .0034.

18.4   19/4,096 or .0046.

18.5   (a) less than .0001; (b) .0287; (c) .5398.

18.6   (a) less than .0001; (b) less than .0001; (c) less than .0001.

19.1   (a) 1.25; (b) 47.94 to 52.06.

19.2   (a) 3.00; (b) .55; (c) 20.93 to 23.07.

20.1 $z = -1.8$ (retain).

20.2   Probability of r equal to or greater than 8 is approximately .0034   (reject).

22.1 $t = -1.50; df = 9$ (retain)
22.2   17.46 to 32.54.
23.1 $t = 1.99; df = 16$ (retain, two-tailed test)
23.2 $t = 2.69; df = 18$ (reject)
23.3 $t = 2.11; df = 57$ (reject)
24.1 $t = 2.00; df = 7$ (reject, one-tailed test)
24.2 $t = 2.31; df = 11$ (reject, one-tailed test)
25.1 $U = 18.5$ (retain, two-tailed test).
25.2 $U = 20.5$ (reject).
25.3 $U = 63$ (reject).
26.1 $T = 3.5$ (retain at .025 level, N reduced to 7).
26.2 $T = 12$ (retain, N reduced to 11).
28.1   Chi-square = 6.26; $df = 1$ (retain, two-tailed probability about .012)
       $z = 2.50, p = .006$ (one-tailed probability)
28.2   Chi-square = 54; $df = 9$ with ten cells   (reject)
28.3   Chi-square = 2.96; $df = 3$ (retain)
29.1   Chi-square = 38.1; $df = 4$ (reject)
29.2   Chi-square = 4.3; $df = 3$ (collapse cells)   (retain)
29.3   Chi-square = 2.13; $df = 1$ (retain)
29.4   Chi-square = 87.1; $df = 9$ (reject) $C = .68$
30.1   Retain.
30.2   Reject.
30.3   Retain.
30.4   Retain.
30.5   Chi-square = 5.12; $df = 1$ (reject).
31.1 $D = .186$ (reject), using 10 intervals.
31.2 $D = .25$ (retain), using 10 intervals; $D = .30$ (retain), using 20 intervals.
32.1 $K_D = 7$ (retain), using 30 intervals.
32.2 $D = .15$ (retain).
33.1   Reject.
33.2   Chi-square = 4.45, $df = 1$ (reject).
35.1 $F = 7.00/.44 = 15.9, df = 2,9$ (reject).
35.2 $F = 6.53/7.04 = .93, df = 3,16$ (retain).
35.3 $F = 17.8/4.5 = 3.95, df = 1,16$ (retain).
36.1 M_1 and M_2 both significantly smaller than M_3; no significant difference between M_1 and M_2.
36.2 M_1 significantly smaller than M_3; no significant difference between M_1 and M_2.
37.1   (a) $F = .20/1.30 = .15, df = 1.16$ (retain); (b) $F = 1.80/1.30 = 1.38, df = 1,16$ (retain); (c) $F = 5.00/1.30 = 3.85, df = 1,16$ (retain).
38.1   (a) $F = 5.00/.28 = 18.0, df = 1,6$ (reject); (b) 4; 5.5.
38.2   (a) $F = 1.80/1.24 = 1.45, df = 1,17$ (retain); (b) 2.2; 1.6.
39.1   (a) $\hat{Y} = .714X_1 + .054X_2 - 3.857$; (b) .846; (c) .878.
39.2   (a) $\hat{Y} = 1.56X_1 + 1.06$; (b) .781; (c) .854; (d) $\hat{Y} = 2.5X - .5X^2 + 1$; (e) .80; (f) 1.00.

INDEX

Guilford

Quantité 2nd° ceme —

Eta 4° ceme

fannummmie

Plus 20% to Sample number

II.

1. Given the scores: 9, 4, 7, 6, 10, 12. Solve for the mean and standard deviation.

2. On a certain achievement test the seventh grade mean is 28.00 and the standard deviation is 4.8, while the eighth grade mean is 31.6 and the standard deviation is 4.00 What per cent of the seventh graders scored above the mean of the eighth grade? What per cent of the eighth graders scored below the mean of the seventh grade?

3. Given a distribution with a mean of 88.6, a standard deviation of 7.5, and N of 626, compute the 95% confidence interval for the population mean.

4. Robert obtained a score of 630 on a test with a mean of 500 and a standard deviation of 100. If the distribution on this test was normal, determine the percentile rank of Robert's score.

5. Given the following scores for two groups of students:

   Group I: 2, 5, 7, 9, 6, 7

   Group II: 4, 16, 11, 9, 8

   Is there a significant difference between the means of the two groups at the .05 level?

6. The mean IQ of tenth grade students was 100 with a standard deviation of 15. The same group obtained a mean of 75 on the mathematics Regents examination with a standard deviation of 12. A correlation of 0.7 was found between IQ and the mathematics Regents examination. Derive the prediction equation to estimate mathematics score from IQ. Estimate the score on the mathematics examination of a student whose IQ is 120.